SEX AT DUSK

Lifting the Shiny Wrapping from

SEX AT DAWN

LYNN SAXON

Copyright © 2012 Lynn Saxon

All rights reserved.

ISBN-10: 1477697284
ISBN-13: 978-1477697283

CONTENTS

Preface   v

INTRODUCTION   1

CHAPTER ONE
Darwinian Natural Selection   11

CHAPTER TWO
The Birds and the Bees for Adults   45

CHAPTER THREE
Apes   77

CHAPTER FOUR
Paternity   111

CHAPTER FIVE
Parenting and Marriage   149

CHAPTER SIX
Monogamy and Jealousy   181

CHAPTER SEVEN
The Way We Were? 211

CHAPTER EIGHT
Body Talk 243

CHAPTER NINE
Let's Hear it for the Girls
(Though mostly it's still about the boys) 275

CHAPTER TEN
What Men Want? 299

Sluts or Whores? 327

REFERENCES 335

INDEX 357

Preface

In the summer of 2010 my attention was drawn to a new book titled *Sex at Dawn* by Christopher Ryan and Cacilda Jethá. Not evolutionary biologists (their interests are psychology and Ms Jethá is a practicing psychiatrist) the authors presented an attack on what they called the 'standard narrative' of evolutionary psychology. This 'standard narrative' argument, they say, is based upon false beliefs about how we lived as modern humans (from about 200,000 years ago), especially the 'false' belief that we lived in sexually monogamous nuclear families with varying degrees of deceitful extra-pair sex.

Ryan and Jethá have presented an alternative argument: we are not naturally monogamous today so our ancestral breeding system must have been one of open multiple simultaneous sexual pairings. They argue that this would not have involved jealousy or deceit or any concern about known paternity but would have been based on sharing: because resources such as food would have been shared in our egalitarian ancestors, so would sex.

While evolutionary psychology is not my main concern, evolutionary biology is. I have read and debated on the evolution of sex and the sexes for many years. Sex is a topic that interests so many of us yet there are few who have any real understanding of the place of sex and the sexes in evolution. When I saw that many of the books referenced by Ryan and Jethá are also on my bookshelves I decided that this popular book deserved to be read so that I might understand how they had reached their conclusions about human sexuality.

The basic argument that we are not naturally sexually monogamous is reasonably sound – anyone involved in evolutionary biology today would not argue against that. (Apart from anything else, to be naturally completely sexually monogamous we would have to be mating for life with our first sexual partner.) But Ryan and Jethá go much further than an argument for serial monogamy or monogamy with some extra-pair sex, and present us as being naturally bonobo-like – bonobos, in their understanding, live in groups where everyone regularly and casually has sex with everyone else.

Reading their book I felt increasingly concerned that their argument was presented as being backed by scientific evidence. There were many blatant errors and false representations which readers were accepting as factual evidence.

The authors have put their book forward as a means to create debate, so here I present evidence which I believe fills in many of their omissions and corrects many of the distortions and errors of their argument. Readers are possibly only interested in *human* sexuality but without a better understanding of the evolutionary biology of sex it is too easy to imagine things about our species which would put us outside of evolution. We're not. Even if we think people today are able to escape 'nature' to a greater or lesser degree, the ancestors we are talking about here, from the common ancestor with the other African apes up to our recent ancestors 10,000 years ago, were naturally evolving in natural environments.

There is still, of course, much to learn about ourselves and our evolution but it would be wrong not to provide people with a chance to understand the evidence so far, even if some of it takes a little work to grasp. I am in the first instance presenting a fuller and corrected picture of the 'evidence' put forward by Ryan and Jethá, and then adding other evidence which has been omitted. While I have included some thoughts and interpretations that are my own, my main intent is simply to get across better and more honest information.

This book is obviously primarily aimed at readers of *Sex at Dawn*. I have also tried to make the arguments of that book as clear as possible so that those who have not read it but are interested in the arguments may be able to follow my response – and understand why it is necessary.

SEX AT DUSK

INTRODUCTION

We are apes. We are animals. Humans – *Homo sapiens* – share a common ancestor with the other African apes: chimpanzee, bonobo, and gorilla. If we trace our ancestry back through time we find that at about five to seven million years ago our ancestor joins with the ancestor of the chimpanzee and the bonobo, and about seven to nine million years ago that common ancestor meets with that of the gorilla.

We cannot know for certain what this common gorilla/chimpanzee/bonobo/human ancestor was like. Was it like a modern gorilla with a male twice the size of the females which he guarded in his little breeding group? Or was it more like the chimpanzee and the bonobo with multimale/multifemale groups and where males and females were much closer in size? And six million years ago at the *Pan/Homo* split (chimpanzees and bonobos are *Pan troglodytes* and *Pan paniscus* respectively), what kind of ape was that common ancestor? We cannot know for sure.

Add to this the various hominin fossils on the line between that common human/chimpanzee/bonobo ancestor to the modern human and we still do not know for sure how the sizes of males and females differed or not through time, or what kinds of social and breeding groups they lived in *en route* to *Homo sapiens*. Yet we look to whatever evidence we *can* find to try to piece together an understanding of our ancestors, often in an attempt to understand more about ourselves today: more about our 'true' nature, whatever that might be. One aspect of this nature we particularly like to debate is our 'natural' sexual nature, and in the modern Western world especially this has become a pressing issue as infidelity and divorce and 'sexual liberation' and 'sexual dysfunction' have changed the world from the one our grandparents knew (or at least thought they did).

Sex – just what is it all about? Don't other species just get on with it? What are 'men' and 'women' and the relationships between them? What about the differences and the conflicts and the jealousies and all the pain and disappointments we so easily cause each other?

In the West in the mid-20th century there was a view of marriage that the eager bride drags the reluctant groom up the aisle. She is believed to be happy to have secured her mate and to be keen to live in monogamous bliss. He is only there because this is his best opportunity for sex on tap; how likely he is to stay monogamous depends only on how many other sexual opportunities come his way. Marriage and 'the wife' are not presented as something he really wanted but something socially expected, and perhaps his only opportunity for sex when women's sexual behaviour was largely constrained by social attitudes. Monogamous marriage was seen, as it often still is (including as noted by Ryan and Jethá, p. 2[1]), as something like a prison sentence for the man, and the wife is his 'ball and chain'.

Then came the women's movement of the sixties and seventies which seemed to offer something a bit different: sexually liberated women who were less keen on being virginal brides and more keen on being like the men; women seeking economic independence and sexual freedoms to lift them from being the 'second sex'[2] and put them on a

---

[1] Page numbers are the same for both editions of *Sex at Dawn*

[2] In reference to Simone de Beauvoir's *The Second Sex*, first published in 1949.

more equal par with the 'first sex' – men. Yet four decades later we still seem none the wiser about sex and relationships. Just what is standing in our way? If the natural world runs so smoothly why is it so hard for us? Just who are the baddies standing in the way of our natural freedom, happiness, and sexual satiation?

Ryan and Jethá[3] in *Sex at Dawn* (p. 2) state that there is good reason to believe that marriage is the beginning and end of a man's sexual life. And though they say that women fare little better, this, they say, is because the wife is left spending her life apologizing for being just one woman! The middle of the book does tell us about the supposedly natural wild promiscuity of the female of the species but this is sandwiched between clear arguments for the greater need for, and greater acceptance of, extramarital sexual encounters for men.

It is hard to imagine that people have ever really believed that the male of the species is 'naturally' sexually monogamous. But he often *does* want a wife, and children he knows to be his. Most human cultures have allowed polygynous marriage – one man with more than one wife – and many have not come down too hard on male infidelity and promiscuity. But for women there has more often been a strict control of their sexual behaviour so that a man could be sure he was only raising his own children. Certain 'unfortunate' women and girls as sex workers, or casual affairs kept going by false promises, have provided the extramarital sexual outlet for men.

Ryan and Jethá's argument is that before about 10,000 years ago sexual constraints did not exist, paternity was hardly an issue, and men and women engaged in fairly free and casual sexual activity. Surely, they argue, our inability to sustain sexually monogamous marriages even with the constraints enforced by laws and religions and moralizers proves that open promiscuity must have been how our ancestors lived.

---

[3] I am using the names of both authors though Christopher Ryan is the writer while his wife, Cacilda Jethá, "read every draft, again and again" and, as a medical doctor, was "integral to the discussions of diet, longevity, infant care, and so on." (Ryan in conversation with Dan Savage in the 2011 paperback edition of *Sex at Dawn*).

While many people *do* still sustain happy and sexually monogamous relationships it is also true that there are many who do not. Is this a clear sign that sexual monogamy is not 'natural' and, as the authors argue, that we are really as promiscuous as *Pan* (chimpanzees and bonobos)? Just why *is* 'sex' such a problem and relationships in such apparent crisis? What does the ubiquitous sex industry tell us about ourselves? Where does this leave the nuclear family unit? Where does it leave children? Has there been some false story-telling about pair-bonding in our evolution which has been used to support a false argument for a natural nuclear family unit going way back to the dawn of humankind? Have we all really been duped?

Ryan and Jethá criticize what they call "the standard narrative of human social evolution". This, they say (p. 7), is the evolutionary psychology argument which says that men and women assess each other's mate value as a reproductive resource, form pair bonds, and then jealously guard the resources they have acquired in each other while each follows a self-interested reproductive strategy of sexual cheating. The argument they are attacking, then, with this extra-pair sexual cheating, is not one of sexual monogamy after all. They are arguing that our ancestors not only were not sexually monogamous, they were also not socially monogamous, i.e., there were no bonded reproductive mates to cheat on. By removing social monogamy they can eliminate the nasty 'cheating' side of human sexual behaviour.

If, they argue, we see our ancestors as fiercely egalitarian, sharing all resources, then there is no need for laying claims to a specific individual as a reproductive resource, and every reason to believe sex itself would have to be shared. And they say they have the evidence for this: evidence in our bodies, in the habits of the remaining relatively isolated societies, and in certain aspects of contemporary Western culture.

In the chapters that follow I will look at the evidence again. Don't expect easy answers as life is always more complex than we imagine – just as we think we have found one answer, new questions are often opened up. Will we find casual sexual behaviour or is there something about the human ape that made that very improbable? Just what does go on in the sex lives of chimpanzees and bonobos – and just how casual and sexy is sex in those bonobos? What clues can hunter-gatherer tribes give us?

The human female, more than the female of any other species, *has* had her sexuality socially controlled. She has often only been able to access resources through a man, exchanging sexual/reproductive access to her body for her access to food, shelter, and protection. How far back might that go? Are Ryan and Jethá correct that this is only a result of the shift to settled, agricultural communities? Or do we have evidence for anything similar in foragers, and even the other apes?

Modern *Homo sapiens*, people much like ourselves, have been around for about 200,000 years. What sorts of lives did these forager ancestors live? If egalitarianism and food sharing was enforced in tribal living do we have reason to believe, as Ryan and Jethá argue, that something similar was enforced in sexual behaviour?

The widespread anthropological agreement – not just that of evolutionary psychology – is that our evolution involved pair-bonding with some polygynous males having more than one wife but most mated men monogamously paired. Obviously with a more or less equal ratio of males and females one man's gain of an extra wife is another man's loss. While women were more likely to be officially mated to only one male there was also some (probably relatively hidden) polyandrous (multiple male) mating by females. Though sex was not 'shared' within groups there may well have been varying degrees of extra-pair sexual liaisons and relatively easy divorce and serial monogamy; divorces were most likely often *due* to those extra-pair sexual liaisons becoming known or intolerable to a spouse.

The big question is: do we have any reason to believe that there was a pre-agricultural promiscuity where men and women mated, often casually and openly, with many others? Does the necessity of the sharing of food and other resources amongst a relatively small group lead to the logical conclusion that mates would be shared too? And if not then why not?

I believe the strongest evidence is that pair-bonding, mate-guarding, and male parental investment in (usually) his own biological offspring extends far back in our evolution, and that casual sex within a group would not and could not have been workable for *Homo sapiens* – it could not have evolved. A norm of multiple open simultaneous pairings by women as well as men, I will argue, did not exist. Agriculture certainly changed a lot for humans but it was a change in degree of traits that were already well established rather than a change in kind.

If we are going to talk about groups and resources and sharing we need to understand what it is we are talking about. It is straightforward to understand what is meant by food and protection as *resources* and why they might be shared, especially if it is reciprocal exchange of like for like. But what do we mean when we talk of *sex* as a resource that can be shared? Just what is this 'resource', and how much of 'it' is there, and is it something that is the same for both sexes and exchanged like for like?

While multiple females may share the same resource of DNA coming from the same male, females are not transferring *their* DNA into males during sex. There is no two-way sharing between the sexes in this sense. The resource that the female provides in terms of maternal input into a single offspring cannot be shared between males. So what is meant by sex as a resource that, like food, can be shared?

Darwin treated sexual selection differently from the survival aspects of natural selection because he saw that its effects were quite different from those connected to survival. We need to understand why sex is different. Is it a need like food? Organisms do not die from the lack of sex – but the future of their genes does. This future for the genes, as we'll see, even turns out to override the survival needs of the organism itself.

In the West we emphasize recreation rather than procreation in connection to our sexual needs, though why this particular recreational activity should have such a powerful control over us is not fully appreciated. We seek and expect the physiological rewards, which is what we *think* it is about, yet we despair at the negative aspects that also spoil the party. If sex is fun why is such misery connected to it? Only by understanding the evolution of sex and the sexes will we understand why this is, and understand how 'recreational' sex is inextricably tied to procreation and reproductive success.

Sex at Dawn constantly reminded me of a line from the novel *Nice Work* by David Lodge (1988):

"Literature is mostly about having sex and not much about having children; life's the other way round."

Sex at Dawn is almost all about sex and not much about children, yet evolution is very much about reproduction – variation in reproductive success *is* evolution.

For most of our history girls would have been having their first sex soon after puberty and, though puberty likely was later than it is today, would have become mothers by the time they reached their twenties. Today we have the luxury of avoiding parenthood altogether or extending the pre-parenthood period of our lives well beyond what our ancestors, especially our female ancestors, experienced. Children are far less a part of our lives than they were in our past and this is especially true for women who no longer have an infant strapped to their body from their late teens. This surely impacts our sexual behaviour.

Ryan and Jethá simply say that all men in the group will all equally share the parenting of the offspring because it is *for the good of the group,* but group selection arguments have been almost totally dismissed in evolutionary biology and with good reason. The unit on which selection acts is the 'gene', and the 'gene's eye' view has given us the best (and sometimes the only) way to understand what we observe across species. If behaviour is about helping a close relative then it is helping shared genes in another body which is kin selection. Beyond this there can also be reciprocal exchange where one person's excess shared on one occasion will be paid back on another occasion. Joint ventures can also provide the best outcome for individuals, and when we understand that the joint venture of sexual reproduction is most often between 'strangers' who do not share genes to the extent that close kin share genes, we can gain some insight into the conflicts of interest that arise.

Genes cannot be selected that sacrifice their own continued existence so that *different* genes can continue instead; self-sacrificing genes have no future. Genes that are good at producing traits that get more copies of themselves into new individuals spread, those that are bad at it don't. When the production of new individuals by females is limited (as it most often is), genes that produce traits in males that make those males good at ensuring their genes are in those offspring are going to be selected over the alternative genes and traits in other males that are less good. A genetic indifference about success at achieving paternity cannot be selected. We will come back to this later as it is important though not always easy to understand or accept, not least because of the confusion between 'selfish genes' and 'selfish individuals'.

We also have to remember that the members of family and social groups are not all closely genetically related individuals but comprise both close genetic relatives and marital/sexual ones. In kinship terms this is the distinction between *consanguine* (blood) relatives and *affinal* (marital) relatives. Thinking of a group as a bounded entity through time (and therefore a unit on which selection can act) is a mistake because individuals move between groups, usually *in order to* breed with non-relatives, and it is this that creates the gene-flow throughout a species; separate, bounded groups without gene flow between them would ultimately evolve into separate species.

So we have to take into consideration the difference between blood relatives and those non-genetic (or less closely genetic, as most genes are shared within a species) relations between people due to marriage and sexual reproduction, and how the membership of a group, both in terms of individuals and of genes, constantly changes through time. The group as the entity on which natural selection acts – its descendants out-numbering those of other groups – does not work because individuals, and therefore any selected genes and selected traits, move between groups.

Some of the things we will be looking at may take a little effort to follow but they are important considerations if we are to avoid simplistic and faulty conclusions.

If we just look at this thing called 'sex' as being a need and a resource on a par with food which is being 'unnaturally' or anti-socially restricted, then removing restrictions so that the resource of sex can be shared seems a straightforward solution. But when we understand what sex is we can understand why it is different from food.

If we are going to look at chimpanzees and bonobos for clues about ourselves then we have to know what is actually going on in chimpanzees and bonobos and not merely take a few observations that fit a particular agenda and ignore everything else.

If we are going to use evolutionary biology in our argument then we have to understand evolutionary biology and not merely dismiss some evolutionary psychology accounts.

And if we are going to talk about what has been naturally and sexually selected then we have to understand natural and sexual selection.

Ryan and Jethá's argument fails on all of these counts and they have substituted one evolutionary psychology story with another that has even less evidential support.

So now let's look at the evidence, fill in some gaps, and make some corrections.

CHAPTER ONE

Darwinian Natural Selection

While evolutionary psychology seeks to explain current human psychology and behaviour by appealing to how we supposedly spent some two hundred thousand years of pre-modern evolution, it should not be confused with the sciences of evolutionary biology and natural selection. Evolutionary psychology is focused on current, usually Western, human behaviours, and it tends to easily lose contact with the science of evolutionary biology after taking in a few simple generalizations about males and females and applying them to all kinds of weird and wonderful things. And this includes the evolutionary psychology story told in *Sex at Dawn*.

In order to get a better understanding of the possible evolutionary influences on human sexual behaviour we first need to get a better understanding of the evolution of sex itself. I will go into this in some detail in this and the next chapter as we travel alongside the criticisms and arguments presented in *Sex at Dawn*.

Ryan and Jethá (p. 27) say they accept that female sexual reticence and choosiness – the 'coy' or discriminating female as described by Darwin – is a key feature in the mating systems of many mammals but they believe that it is not particularly applicable to humans or to the primates most closely related to us. Ryan and Jethá see 'the standard narrative' evolutionary psychology as an incorrect application of natural selection to humans but their mockery of Darwin and natural selection in respect of human sexuality is not obviously distinguishable from an attack on natural selection itself.

The absence of anything much in the way of an explanation of general evolutionary theory in *Sex at Dawn* means that their evolutionary story only really begins with bonobos and chimpanzees, quickly leaping to modern humans from 200,000 years ago, and therefore comprises only the failure, in their view, of evolutionary theory. Though the authors may accept that evolutionary theory does apply to other species there is no indication that they understand why this is so, i.e., why evolutionary theory and especially natural selection are such powerful concepts.

If Darwinian natural selection is to be mocked with regard to humans then we should at least start with some understanding of it as it does apply to other species. We cannot make a special case for human evolution without understanding something of the evolutionary mechanisms that have been in action for hundreds of millions of years in all our ancestors through all that time. We cannot, as religions do, make a special case for humans simply because we want it to be so.

Darwin and natural selection

Charles Darwin was a Victorian gentleman and his views on the two sexes reflected those of his time and social position. His interest was not so much in people specifically as in the whole of evolution and how different species evolved from a common ancestor. His 1859 book *On the Origin of Species* is his theory of how evolution works via the mechanism of natural selection: traits spread in a population down through generations because they aid survival in particular environments better than the alternative traits. But this left Darwin with a problem: many males have traits that hinder rather than help their survival.

To deal with this problem Darwin then went on to write *The Descent of Man, and Selection in Relation to Sex* (first published in 1871) about sexual selection, explaining the competition between males for females and female choosiness. While most people at the time could accept male-male competition and the selection of traits for vigour and strength in males, female choice was not accepted and, floundering under the arguments, it would be a hundred years before active female mate choice would be taken seriously. The 'coy' female who passively accepts the winning male was as far as female sexual behaviour could be viewed to go for it to be acceptable, and even Darwin's more active female choice of aesthetically pleasing traits in males was too much for his contemporaries to handle (Cronin 1991).

The recognition that females may actually choose to mate with more than one male would be a long time coming, and so it is not surprising that Darwin does not seem to have considered wilful polyandrous mating by females, though he certainly was aware of it. In some of the barnacle species he studied he noted that he found a number of dwarf males inside the female. Writing to his friend, Charles Lyell, he described how the female in two of the valves of her shell:

"...had two little pockets, in each of which she kept a little husband; I do not know of any other case where a female invariably has two husbands..."

Darwin described these males as little more than bags of spermatozoa, and he went on to discover other barnacles with as many as 14 miniature males in a single female, speculating that any one of them might fertilize the female's eggs (Birkhead 2000).

Females of many species are now known to mate more often than is necessary for the fertilization of their eggs, and females will also mate with multiple males rather than choosing only the best male. Discovering why this is so is an interesting and important topic in evolutionary biology, not least because mating is actually quite a costly endeavour in the natural world. These costs include reduced time for foraging, energy used in finding a mate, increased risks of predation, risks of injury in competition and even during mating itself, and risks of disease transmission; as we will see, mating is often resisted by the egg producers (and sometimes by the sperm producers). If this female resistance has been *selected against* in a species then it requires an explanation as to why.

Natural selection is the mechanism by which evolution occurs due to certain traits being selected over their alternatives. Traits that better enable survival and reproduction in a particular environment are the ones that, on average, are passed on to more offspring. Darwin did not know about genes and it would be many decades before they were discovered as the carriers of the recipe for development, and recognized as the means by which inheritance of traits occurs.

Evolution is now defined as *the change in the frequency of genes*, or *alleles* to be more precise – alleles are the different versions of the same genes, such as those that produce the different variants of eye colour. Natural selection is a mechanism of evolution which is recognized as being in action when particular traits spread down through the generations because they provide survival and reproductive advantages over the alternative traits. Evolution can, and often does, occur without natural selection: the different proportions of genes for different eye colours, for example, may vary down through generations without there being any survival or reproductive advantage that is being directly selected.

Serving eggs, serving sperm

What would become the two sexes began with the evolution of two types of sex cell: males (almost always) produce vast numbers of tiny sperm and can (potentially) make vast numbers of offspring, while females produce fewer and larger eggs. In most species most of the available eggs will be fertilized and most sperm will fail to fertilize. Being an egg producer *or* a sperm producer is how female and male came into being and it is how the sexes are defined in biology.

However much we as humans argue about and redefine 'gender', in the evolution of sex the two distinct sex cell types remain a fact that has enormous consequences in natural selection: two different – and sometimes very different – naturally selected[4] outcomes are found *within the same species*. This is a basic fact about sexually reproducing species which cannot be ignored. For the male of most species the number of copies of his genes that he passes on is limited less by his sperm production than by the limited number of available eggs and females. These eggs and females do not treat all sperm, or males, as equally desirable, and we should also note that males do not necessarily find all females equally desirable either.

When we look at most species, especially those that are not monogamous (and strict sexual monogamy is rare), we see that males and females can easily be distinguished. This is known as *sexual dimorphism*. In some species it is far from obvious that males and females are even of the same species; this was the case with the mallard duck which the father of taxonomy, Carl Linnaeus (1707-1778), first classified as different species (Andersson 1994).

In some species, such as those barnacles Darwin studied, males are reduced to packets of sperm and live on or in the female. In many species we have brightly coloured males and drab females, or males

---

[4] "Natural selection" is normally used to include "sexual selection", the latter term, though, is often necessary in contexts where selection due to sex needs to be clearly distinguished from that due to survival.

with horns or antlers, or males twice the size, or a fraction of the size, of females.

Why?

It is because selection has acted differently on the two sexes.

Why?

Because in sexual selection genes spread when an individual out-reproduces others *of the same sex*. Selection is acting differently on the two sexes even though they are of the same species with the same genome. The physical and behavioural differences that arise between the two sexes of the same species are differences that serve the reproductive interests of their respective sex cells. Males have their male-specific traits because those traits help genes that are in sperm get into more descendants, females have their female-specific traits because those traits help genes that are in eggs get into more descendants.

It is the variation between the male bodies and behaviours that sexual selection acts upon, and the same, separately, with variations between the female bodies and behaviours. Selection pressures can be quite different on the two sexes with regard to reproductive success so the two sexes, while sharing the same genome, diverge in the traits that they express.

In deep sea angler fish species, for example, only the female becomes a full adult; the male larva develops to become little more than sensory organs which are needed in order to find a female. If the male does encounter a female he then bites into her body and melds with it, reducing to little more than male gonads attached to her body. There can be multiple males attached to one female in this way, kept alive by the female and providing sperm for her eggs when needed (Andersson 1994).

Many marine species have reduced males like this but we are more used to thinking about mammals, and thinking of males as being bigger and stronger than females and physically battling for status and ultimately sexual access to females. Most species, though, have smaller males than females often because a bigger female produces more eggs while the male can produce his sperm cells and remain small.

Males produce sperm, females produce eggs, and so traits that lead to greater reproductive success among sperm producers get passed on to more male offspring than do alternative traits, and traits that lead to greater reproductive success among egg producers get passed on to more female offspring. To put it more correctly, the traits are expressed in one sex or the other, as both sexes carry the same genes.[5] With the evolution of two sexes, mechanisms such as sex hormones evolved to limit the expression of sexually selected traits to one sex, though this is not always fully achieved.

These basic facts about evolution, and especially about sexual selection, need to be grasped in order for us to make proper suggestions and hypotheses about the evolution of any sexually reproducing species past or present. Evolutionary psychology may well be mocked and ridiculed, as Ryan and Jethá do (and can themselves be), for presenting unsubstantiated assertions about human adaptations for one thing or another but natural selection and sexual selection cannot. There are arguments about the details, and arguments as to whether selection has in fact happened or whether it is chance or some other non-selective mechanism that has led to a particular trait, but the foundations of natural selection are sound.

For humans we are also faced with the difficulty of being objective about ourselves and without self-interested agendas, and culture of course is also clearly highly relevant. But if we accept that we have evolved and we are going to use natural selection to understand ourselves or our ancestors, as Ryan and Jethá say they are, we need to take the science of evolutionary biology seriously.

Ryan and Jethá are correct when they say we have learned a lot more about animal sexual behaviour since Darwin, but in mocking Darwin in this respect the authors have failed to acknowledge that he did at least see the existence and importance of female mate choice and

---

[5] Except for a very small number in some species that are on a sex chromosome, such as those on the Y chromosome in mammals.

held on to this belief in spite of much opposition. Darwin argued for an active female role in evolution beyond the passive one argued for by his contemporaries; as we shall see, a passive female role is also the outcome of Ryan and Jethá's argument for casual female promiscuity.

Female mate choice was finally, in the 1970s, given the full attention Darwin would have wanted. (This was also a time in the West of an active and significant women's movement which may well have some connection.) As well as looking at how female mate choice led to display ornaments in males, such as showy feathers and bright colours, it also became obvious that females were often mating with more males than (from previous thinking) they needed to or, it was believed, did.

Recognizing that females in many species are not essentially monogamous has been an immensely important breakthrough but 'why' still needs to be understood. To argue, as Ryan and Jethá ultimately do[6], that it is because females are like males and simply want and enjoy sex in much the same way as do males, is really not good enough. This argument emanates partly from the desire for sexual equality that relies upon revealing women to be the same as men when traditionally that could only be seen as raising the status of women. While we do struggle more to unveil 'natural' or unconstrained female sexuality than male sexuality it would nevertheless be a mistake – and no less sexist – to assume that the sexes are both essentially 'male'.

We now need to delve more deeply and to look at some of the people and ideas associated with evolutionary biology that are (mostly) mocked and dismissed by Ryan and Jethá.

Angus Bateman

Back in 1948 geneticist Angus Bateman carried out some experiments on fruit flies which at the time helped to establish empirical support for

---

[6] Most directly in the postscript to the paperback edition.

the promiscuous male/choosy female dichotomy. His results from a series of experiments showed male fruit flies increasing their number of offspring by mating with more females, while the females could not increase their number of offspring by mating with more males. These experiments are now known not to be as clear-cut as previously thought. The full results show that females produced progeny fathered by more than one male (though not necessarily a result of female 'choice') and in most cases female offspring numbers did increase (though less so than that of some of the males) when females mated with more males.

The species used, *Drosophila melanogaster*, is now known to have females able to store sperm for 3-4 days whereas in other fruit fly species the females cannot store sperm and they need to mate more often (in *D. pseudoobscura* every few hours) to acquire enough. If Bateman's experiments had been carried out with a different species of fruit fly or had gone on for longer than the three or four days they did, then females would have been seen to be keen to remate. It has also been discovered that the male fruit fly's seminal fluid contains chemicals which act as an anti-aphrodisiac to delay the female's interest in mating with other males (Birkhead 2000, Tang-Martinez and Brandt Ryder 2005).

Because males – and this applies to all species – can potentially fertilize the eggs of a number of different females whereas the females have a limited number of eggs, the *variance* in the numbers of offspring from males is often greater than the variance in offspring numbers in females. In the 64 experiments by Bateman, 21% of males *failed* to fertilize any eggs while only 4% of the females failed to produce young (Hrdy 1986), so this does show the *greater variance* in male reproductive success compared to female which, Bateman said, was what *determined the greater eagerness of males to mate*. The males who successfully compete to fertilize more of the available eggs will pass their genes on to larger numbers of offspring while those who fail in the competition will pass theirs to few or none.

Ryan and Jethá (p. 52) say only that Angus Bateman "wasn't hesitant to extrapolate his findings concerning fruit fly behavior to humans" when Bateman wrote that natural selection encourages "an indiscriminating eagerness in the males and a discriminating passivity in the females."

To put this in context Bateman actually wrote (note that gametes are the eggs and sperm):

"The primary feature of sexual selection is to be sure the fusion of gametes irrespective of their relative size, but the specialization into large immobile gamete and small mobile gametes produced in great excess (the primary sex difference), was a very early evolutionary step. One would therefore expect to find in all but a few very primitive organisms, and those in which monogamy combined with a sex ratio of unity eliminated all intra-sexual selection, that males would show greater intra-sexual selection than females. This would explain why in unisexual organisms there is always a combination of an indiscriminating eagerness in the males and a discriminating passivity in the females. Even in derived monogamous species (e.g., man) this sex difference might be expected to persist as a relic." (Bateman 1948)

Not exactly the sexist eagerness by Bateman to apply his findings to humans that is suggested by Ryan and Jethá.

It is hardly going against evolutionary science to look for evolutionary principles that apply across species, including humans. Any non-religious, evolutionary based 'special case' for humans will still need to be supported by evidence for evolved mechanisms able to produce 'special case' human traits. Evolutionary biology has moved on a lot since Bateman but there continues to be much about his work that remains relevant, and Bateman's principles are still debated and potentially useful as, for example, the title of one paper: "Don't throw Bateman out with the bathwater" (Wade and Schuster 2005) suggests.

Bateman's basic insight that the greater variance in the reproductive output of one sex compared to the other leads to a greater eagerness to mate by that sex still stands, only rather than it being just the males who have the greater variance we now know that in some species it is the females. The sex – male *or* female – which has the greater variance in numbers of offspring is the sex most eager to mate. In species where the sex roles are reversed and the females have the greater variance in numbers of offspring it is the females who are most eager to mate and the males who are the choosy sex, as happens, for example, in phalarope shorebirds (Reynolds, Colwell, and Cooke 1986) as we'll see below, and pipefish (Jones, Walker, and Avise 2001).

Robert Trivers

This brings in another important aspect of reproduction besides the mating: parenting. It is often when one sex is busy parenting a brood while the other sex has the eggs or sperm available for fertilization that the differences in eagerness to mate exist. It was Robert Trivers who went on to show that it was *parental investment* rather than the sex of the individual that was the important factor. The sex with the largest parental investment is a *limiting resource* for which members of the other sex compete (Trivers1972).

So here we have the use of the word *resource* as it applies to sex and reproduction. Males and females are not offering the same resources in equal measure. When the male is the one with the greater parenting resource, such as in the sex role reversed phalarope species of shorebirds, the females are larger and more aggressive and compete for the males. A female can produce a clutch of eggs for one male, and while he is incubating those eggs she can produce a second clutch for another male and even a third and a fourth. The resource we are talking about – parental investment – is a reproductive resource. For most species there are never enough eggs or there is never enough female parenting capacity for the number of available sperm, and in the case of species such as phalaropes there is never enough male parenting capacity for the available numbers of females and their eggs.

When females are removed from the 'mating market', such as in mammals due to the conception, gestation, and milk provisioning that falls exclusively to the female, the ratio of reproductively available males to females, known as the *operational sex ratio*, is often male-biased: there are many more males seeking females for their eggs than females seeking males for their sperm. Females are not expected to vary much in their number of offspring (the variance between females in reproductive success, and therefore female-female competition, is unfortunately a much neglected area for study) while males can potentially succeed in fertilizing the eggs of many females – *or none*.

Geoff Parker and sperm competition

Then something else came from the new look at sexual selection in the 1970s. Geoff Parker studied dung flies and specifically male-male competition. He watched as males would mate with a female only to be forced off by another male who then mated and so on. Females were routinely mating with multiple males and Parker realized that competition between males could continue between the sperm from different males within the female. With this realization he laid the foundations for the study of sperm competition, though Parker was still concerned with the competition between males, and the females were still viewed as the passive ground on which male-male competition played out.

But then in bird studies in the 1980s socially monogamous females were discovered to be actively seeking matings with more than one male so female mate choice became of interest again, this time including 'cryptic female choice' where the choice is made internally between different sperm that are potentially from different males (Birkhead 2000).

The selfish gene

I'll return to sperm competition later but another shift in our understanding of natural and sexual selection also took place from about the 1970s. This was a shift away from seeing selection acting for the benefit of the group or the species to acting for the benefit of the individual and then, sometimes at odds with the individual, more specifically for the benefit of the gene (Dawkins 1976).

Evolution, as already stated, is the change in gene frequencies, with genes that are better at getting themselves into more offspring spreading in subsequent generations. Of course, there is no conscious intention involved, just a logical outcome that genes get more copies of themselves into more offspring if they are producing traits that make them more successful than the alternatives at being reproduced. This means that the well-being of the group, or the species, is certainly not of 'concern' to the gene but neither, necessarily, is the long-term well-being of the individual body in which they currently reside.

Perhaps this is best illustrated by the Australian mouse-like marsupial *antechinus* in which the male has a single mating season of two weeks at the end of his short (11.5 months) life. When it comes to the mating season this little male stops eating and frantically seeks females for sex. His digestive system breaks down and his levels of corticosteroids (stress hormones) skyrocket and his immune system fails. By the end of two weeks he is emaciated, ulcerated, infested, and completely physiologically beat – and dead. All his energy and focus has been devoted to competing with other males, mating (if successful), and even mate-guarding females with copulations lasting up to 12 hours (Lee and Cockburn 1985, Zuk 2007).

For the antechinus, male-male competition has led to this frantic intensive mating period at the end of which he simply drops dead. Females may live and breed for another year or two; obviously such frantic sexual behaviour in a female mammal would leave her genes with no future. But for the male the genes for this behaviour have been selected because they did better at getting passed on than their alternatives. So much for the individual!

To a lesser degree males of many species have evolved bodies and behaviours that lead to a shorter life expectancy than the female of the species. Some receive injury to their bodies, sometimes fatal, in their competition for mates. Those with the bright and sometimes cumbersome ornaments are at greater risk of predation. Others are sometimes eaten by the female during mating: males in many species try to avoid this fate but at least one, the redback spider, intentionally flips his body into a position above the jaws of the female in order to be eaten during mating. Redback spider males only have about a one in five chance of finding a female and if she feeds on the male at the same time as mating she will mate for longer so the male is then able to transfer more sperm and fertilize more eggs. Somewhere in their ancestry a male had a trait for this behaviour, copulated longer, and left more offspring also with this trait so its frequency spread in subsequent generations (Andrade 1996).

Selection has also led to some mothers becoming food for their offspring as in some spiders (Elgar 2005). Birds sometimes kill their siblings (Edwards and Collopy 1983), as do some sharks before they are even born (Prager 2011). New genes and gene variants arise, and if the traits they produce lead to greater survival and reproductive success than alternative traits then they spread.

Though the actual consumption of the mother by some spiders is an extreme example of the costs to the individual in providing a future for their genes, we can also consider more general costs involved in the maternal care of offspring. Why should an individual put so much time and effort into the well-being of resource-hungry other individuals that come out of her body? Why evolve such costly-to-the-self behaviour? Parental behaviour is often seen as the epitome of selfless behaviour but this selfless, individually costly behaviour has been selected for by the success of 'selfish genes'.

We tend to overlook how 'selfish gene' natural selection has acted to produce behaviours such as mothering. Often we simply admire maternal sacrifice as a given, 'selfless' behaviour rather than recognizing it for what it really is: behaviour that benefits the future of the genes while carrying a (sometimes heavy) cost for *the individual* that is the mother.

If genes produce traits that lead to more offspring carrying those genes than carry the alternatives then their numbers increase. There is no greater plan going on, no concern about the future, and no ultimate goal.

Sexual conflict

There is now one more aspect of sexual selection that has become increasingly visible: sexual conflict. The joint project of sexual reproduction often hides the existence of conflict between male and female, as if the fact that each sex needs the other in order to reproduce means that their reproductive interests must therefore converge. As it is usually the males who potentially have more to gain from any one act of copulation (fertilizing eggs with possibly no further costs) and more to lose from not mating (missing perhaps the only fertilization opportunity that is coming his way), males have been noted to evolve sexual 'persistence' traits and females, in response, to evolve 'resistance' traits.

In a more general sexual selection context we can see this in, for example, the ornaments that evolve to greater and greater flamboyance in the attempt to persuade a female to mate. But there are also traits that evolve in males to achieve matings or fertilizations that are more obviously circumventing female choice and can even actually harm the female's lifetime reproductive fitness. Sexual conflict is basically the conflict between the sexes over if, when, and how often to mate because the two sexes often have different naturally selected optima in these respects because of the different reproductive fitness outcomes that result.

Earlier I mentioned the fruit fly semen containing a chemical that acts as an anti-aphrodisiac on the female. The semen also has chemicals that increase the egg production rate of the female, and others which have evolved to incapacitate the sperm of rival males. These chemicals have evolved in the male for his reproductive benefit and not that of the female: the 'sperm competition' chemicals are similar to spider venom, are toxic to the female and can reduce her lifespan (Birkhead 2000).

The male fruit fly's reproductive interest in a female is only in that single reproductive event with that female, so her reduced lifespan is of no consequence to him for there will be no future joint reproductive endeavour. Being able to increase his reproductive output with that female, even if it harms her by reducing her future reproductive output (which would be with other males), does not harm *his* genes but benefits them. Those genes and traits have been selected in the male.

Experiments have been carried out with fruit flies where the females were stopped from co-evolving with males and therefore were prevented from evolving counter-measures against the toxicity of male semen evolving for male-male sperm competition. In these experiments males evolved with semen which was increasingly damaging to females (Rice 1996).

Then experiments were done where monogamy was enforced in fruit flies, and now that the reproductive interests of male and female converged on the same offspring and male sperm competition was removed, the semen became less and less damaging to the females. In addition to this, males also evolved to be less aggressive in their courtship, and reproductive output actually increased over what it had been *with* the sperm competition (Holland and Rice 1999, Pitnick et al. 2001).

Sexual selection involves competition between individuals of the same sex to gain a greater genetic proportional representation in the next generation. If the mate's reproductive success is harmed in the process then selection will not necessarily immediately or completely be able to act to counter that harm. Only with lifetime sexual monogamy can the interests of both parents converge on the very same offspring and extend together for both the parents' lifetimes so that what harms the reproductive fitness of one sex harms that of the other too and is therefore not selected.

When traits in one sex do harm the reproductive fitness of the other sex and there is then selection in that sex to counter at least some of that harm it is known as *sexually antagonistic co-evolution*. It is similar to the co-evolution of predators and prey, or parasites and hosts, and is often described as an 'arms race': an advantage to the male leads to selection on the female to counter that advantage which leads to selection on the male again and so on.

Perhaps it is not how we would prefer to think of sex, reproduction, and the relations between the sexes but evolution is about the differential success of different genes and the traits they produce, not something that exists to please us.

Hermaphrodites

We can even see sexual conflict in hermaphrodites. Most hermaphrodites are invertebrates such as slugs, worms, snails, and flatworms. They are producing both sperm and eggs so we might expect them to just pair up and exchange their sex cells with little fuss – surely no battle of the sexes here? Actually we can see it even more because the 'male side' wants to mate so the 'female side' cannot just get away.

One of the most well-known conflicts in mating is in marine flatworms such as *Pseudobiceros bedfordi* (see overleaf) where the two individuals will 'penis fence' (*P. bedfordi* has two penises).

These duels can last for twenty minutes or more, each one trying to strike the other while avoiding being struck itself. A successful strike deposits sperm on the other which dissolves through the skin leaving it scarred. Flatworms have even been seen that have lost part of the body because the ejaculate has burned right through, though they are able to regenerate (Arnqvist and Rowe 2005).

Two specimens of *Pseudobiceros bedfordi*
about to engage in penis fencing.
(Photo: Nico Michiels)

Garden snails (also hermaphrodites) produce what has been called a 'love dart': a hard structure about 1cm or so in length and made of calcium carbonate. At the end of courtship a 'love dart' is sometimes forced into the body of the mate and then copulation proceeds. This dart introduces mucus into the mate which contains chemicals that make the reproductive system of the mate use the sperm for fertilization of eggs rather than digesting it as food; as little as 0.1% of delivered sperm escapes digestion if a darting has not been successfully achieved. 'Love darts' are not necessary for mating and are not always produced, and even when they are produced they fail to successfully reach their target more than 50% of the time (Rogers and Chase 2001, Arnqvist and Rowe 2005).

There are also a few fish that are hermaphrodites. In hamlet fish and harlequin bass, fertilization is external by spawning and each of the pair takes turns in releasing eggs and releasing sperm. Only a few eggs are released at a time and this is known as egg-trading. If all the eggs were released at once the mate could just release sperm and then leave without providing eggs for the other to fertilize. This egg-trading is what we would expect when, because more offspring can be produced via sperm than via eggs, the male role is the preferred sex role. These species are more likely to cheat by not providing eggs for fertilization because eggs are a more limited and costly resource to produce so they are choosier about releasing them for fertilization.

There are, though, species where the female role seems to be the preferred role. The sea slug *Navanax inermis* also takes turns at being 'female' and 'male', though in this case fertilization is internal and it is cheating on the provision of sperm which is more likely. In *Navanax* the individual that is most keen to mate provides sperm first, suggesting that it is the sperm that is the more valuable resource and is therefore used to encourage mating. In this species sperm can be stored *and* it can be digested, so the receiver has control over whether it is eventually used for fertilization or used as a food source (Leonard and Lukowiak 1991, Leonard 2005).

The production of sperm and the male role were thought automatically to be preferred because of the potentially high number of offspring that can be produced that way, but there is now more evidence and understanding for an advantage in producing eggs. In the female role there can be more control over fertilization and sometimes sperm is used as a food and digested. Egg producers also have more certainty of fertilization due to their limited availability compared to sperm, and sperm producers are always faced with the possibility of no fertilizations.

These hermaphrodites and fruit flies can seem a long way from humans, which indeed they are, but our understanding of many of these natural selection mechanisms can only first be obtained with species that can be observed over many generations in a laboratory (though we also need to consider the differences between laboratory conditions and natural populations). What happens in these species is due to natural selection and it shows how there often is natural conflict between the reproductive interests of the two sexes, or sex cells. It is basically a conflict over the control of fertilization, i.e., that most important of events in natural selection as it concerns the success or failure of genes to exist in the next generation.

Ryan and Jethá's arguments about human sexuality are about a natural lack of conflict between the reproductive interests of the two sexes but they present that argument without even attempting to show any understanding of the discoveries about how sexual conflict does act in evolution. Readers of *Sex at Dawn* who have little or no previous understanding of evolution and natural selection are unlikely to question arguments about the evolution of human sexuality when only human evolution is addressed, albeit alongside some sparse information from other apes. But taking humans out of the whole of the evolutionary adventure to stand alone easily disconnects us from the rest of life, where we have come from, and all the evolutionary mechanisms that potentially apply, or have applied, to us. This, in effect, makes any such argument little different from any other origins myth, including religious ones.

~~~~~~~~~~~~~~~

It is relatively easy to understand why a male might be keen to mate with pretty much any fertile-looking female because selection has led to that being a successful strategy in the males of most species. When bodies have evolved to serve eggs things are a bit different, and female sexual motivation and strategies are often less obvious and more complex and therefore more difficult to elicit. When parenting is involved then selection acts differently again, and we are a species where parenting looms large.

Darwin did not know most of what we now do about sex but he did establish the basic understanding of natural selection and he brought to the fore the importance of sexual selection. We have come a long way from the promiscuous male and the 'coy' female but some, including the authors of *Sex at Dawn*, have been so struck by the fact that females in many species show something other than a monogamous passivity that they have become a little over excited and have sought to show female sexuality as being as indiscriminate as it can often be in the male.

We also tend to overlook male choosiness which also occurs in nature especially when males provide more than just sperm. Even if providing only sperm, males can be choosy in its allocation due to potential sperm depletion (Hardling, Gosden, and Aguilee 2008). Though sperm in terms of how much is potentially available to any one female from the many ready and willing males can be viewed as virtually unlimited, sperm production by any one male *can* be depleted. This is not something that the authors of *Sex at Dawn* deal with, preferring instead to give the impression that, at least in our pre-agricultural past, men naturally had an unlimited supply. And while Ryan and Jethá never directly mention male mate preferences there is much within their book that suggests that the human male, given the choice, would be focusing his sexual attention on young and fertile female bodies; this implicit *male* choosiness runs beneath their explicit argument that human females are naturally not choosy at all.

Darwin did not see a war between the sexes; he viewed species as having either an active male-male competition for a passive female receptivity or a male-male competitive display with active female aesthetic choice of some or other preferred male trait. Whatever cultures have done to shape or control human sexual behaviour, what the *natural* sexual behaviour of other species shows us (and we'll be looking at it more) is that beneath the apparently cooperative reproductive endeavour there is invariably more unpleasantness than we would like to see connected to sex.

Whatever we may feel about the 'unnaturalness' of human marriage and monogamy, nature will not give us an alternative happy ending to the story of sex – there is in reality an unfortunate clash between what we might *wish* 'natural' sex to be and what it in fact is. When something is experienced as a natural and imperative pleasure it is hardly surprising that it leads to the creation of stories which depict some mythological natural world where sex is free and easy, the believers arguing that any constraints on their own pleasure must be due to 'unnatural' forces.

Women have rightly fought against the false representation of female sexuality as invariably passive, reluctant, and 'coy'; it is women who have often been at the forefront of destroying this myth. Active female mate choice is a better representation of the reproductive self-interests of females and it will serve us well to understand the evolutionary biology behind the differences between the sexes. If females of some species evolve to mate more often than do females of other species then it is a valid question to ask why and to look at the different selection pressures experienced. A simplistic conclusion that females are just like males who just like sex does not fit with what we know about evolution and natural selection; it is as much male-biased thinking as anything produced by the Victorians.

When men latch onto the fact that women are not naturally monogamous we should not be surprised if some of them then pursue an agenda that encourages a belief that it is natural for females to simply want sex in the way men do. From the perspective of 'selfish genes' in male bodies there is an obvious benefit to be had from convincing multiple females that it is unnatural to refuse sex, perhaps not far removed from the young (and not so young) men trying to persuade their reluctant young girlfriends to have sex because it is 'only natural'.

Regardless of evolutionary psychology, evolutionary biology is absolutely necessary for our understanding of the very real conflicts of interest that have evolved between the sexes, and the extent that the male will sometimes go to in order to access more of those eggs. In other species the males have no conscious awareness that their evolved behaviours are ultimately connected to accessing eggs and their own reproductive success; just because humans *do* know the connection between sexual behaviours and reproduction (and often deliberately avoid the latter) does not mean that sexual behaviours have dropped, or can drop, their evolved reproductive fitness baggage any more than a decision not to reproduce means we can drop sex.

The apparent removal of reproduction from the equation of sex does not equal the removal of the evolutionary history of 'reproductive fitness' selection for sexual behaviours in the two sexes. Perhaps we also ought to be especially cautious when the reproductive fitness baggage that is being argued should be dropped is only that of the female, leaving us all with just the male sexual promiscuity strategy (i.e., one giant piece of evolutionary fitness baggage carried by the male) for both sexes.

There was certainly a lot Darwin didn't know and couldn't know in his time but it would be wrong to come away from *Sex at Dawn* thinking that evolutionary biology has now discovered female sexuality and mate choice to be on a par with male sexuality and mate choice, or that sexual selection is unimportant or irrelevant with regard to sexual differences. Sexual selection, on the contrary, is turning out to be even more important and more relevant to our understanding of evolution.

Ryan and Jethá (p. 31) take great pleasure in mocking Darwin for using the voyage on the *Beagle* to study the bodies of other species rather than the bodies of the local young human females, saying, for example, that Darwin was too inhibited "to sample the dusky South Pacific pleasures that had inspired the sexually frustrated crew of *The Bounty* to mutiny". Perhaps we should just take a moment to ponder what happened to some of those men whose mutiny was, no doubt, partly inspired by the attractions of living on a South Pacific island but mostly was due to the intolerably oppressive nature of *The Bounty*'s Captain Bligh.

In 1790 nine of the mutineers landed on Pitcairn Island with six male and twelve female Polynesians. That's fifteen men and twelve women. When the colony was discovered eighteen years later there were ten of those women left but only one man (and numerous children). One of the men had committed suicide, one had died, and twelve had been murdered (Ridley 1994). After an initial peaceful four year period the community had fallen into turmoil – the disputes between the men over the women had resulted in the violent deaths of most of these men[7].

Darwin may well have been a buttoned-up Victorian gentleman but at least he laid the foundations for our understanding of natural and sexual selection – and therefore why it wasn't ten men with one "dusky pleasure" left standing.

Thomas Malthus

Ryan and Jethá have a particular concern about Darwin's use of the work of Thomas Malthus. Most of us would support an argument against the inevitability and acceptance of poverty and starvation in the modern world but as far as natural selection goes – and therefore Darwin's use of Malthus – differential survival and differential reproduction *are* the mechanisms of evolutionary change.

For a sexually reproducing population to replace itself it needs only for each female to produce two offspring. So when we see flowers and trees and some animals such as oysters producing thousands, and sometimes millions, of offspring, and even mammals producing well above two per female, it is clear from actual numbers of adults that a minority of those born become parents themselves. Some become food for other species or die from infections or injuries or lack of resources, or simply fail to successfully reproduce. Some females produce fewer

---

[7] http://www.onlinepitcairn.com/history.htm

offspring than others because they get less or poorer quality food than the other females. As we'll see later, getting access to more or better quality food is behind the evolution of social and sexual behaviours in female chimpanzees, bonobos, and humans.

Because all the individuals in a population vary and there are never enough resources to let numbers increase indefinitely (if at all) there is variation in successful access to those resources. Variants of traits that enable some to out-survive and out-reproduce others get passed on to more offspring in the next generation. Frequencies of the different gene variants change: evolution happens. Some variants will be better against parasites or against predators or prey, others may be better at exploiting a new resource or a new or changing environment.

This was what Darwin took from the writings of Malthus which provided the key to understanding evolution – *differential survival and differential reproductive success amongst plants and animals.* This is not a political argument but simply how evolution is; a fact. It does not exclude cooperation, at least between kin or for mutual benefit, but traits that are worse at fighting disease or avoiding predation or acquiring food or successfully reproducing compared to their alternatives are not going to be the ones that, on average, make it through. And it is important to note that just surviving is not enough: as far as evolution is concerned, genes in a successful survivor might as well be in one that was never born if they do not get passed to offspring.

Bodies and behaviours

Selected traits are both physical and behavioural; the two aspects are, after all, linked. Ryan and Jethá write that until E. O. Wilson wrote *Sociobiology* in 1975 evolutionary theory was only interested in how *bodies* came to be as they are. This is because they are again restricting themselves only to thinking about humans, and *Sociobiology*, though predominantly about the social behaviour of other animals, is the book that brought the inheritance of *human* social behaviour into the spotlight.

But animal behaviour had been an area of study for some time: ethologists such as Niko Tinbergen and Konrad Lorenz studied how animal behaviours evolved, and Darwin had also concerned himself with the natural selection of behaviours in his book *The Expression of the Emotions in Man and Animals* (1872). The human part of sociobiology became evolutionary psychology, but an interest in the natural selection of animal behaviour was not itself new.

Bodies *and* behaviours have both genetic and environmental input, and that includes human bodies and behaviours. Natural selection and sexual selection are robust theories with vast amounts of supporting evidence which needs to be incorporated into any theories we may wish to present about ourselves and our past. Reproductive behaviour is no trivial matter in evolution; neither, therefore, is sex.

So, can sex really have been, or become, no more than a casual pastime for any evolved species? Was it ever so for our ancestors?

Lewis Henry Morgan

Near the end of their CHAPTER TWO Ryan and Jethá (p. 42) attack evolutionary psychology for being founded on "the belief that male and female approaches to mating have intrinsically conflicted agendas" but they omit to say that evolutionary biology provides plenty of evidence for these conflicted agendas across species.

In opposition to Darwin's sexual selection they introduce his contemporary Lewis Henry Morgan's (1818-1881) hypothesis that promiscuity and group marriage simply had to be the original forms of our prehistoric social and breeding systems. Without providing any details of this hypothesis or any explanation as to why Morgan believed this to be so, Ryan and Jethá prefer instead to give the impression that Darwin himself was close to agreeing with Morgan.

They state earlier (p. 30): "If you're a Darwin–basher looking for support, you'll find little here", so after knocking and mocking Darwin so much it is as if he then needs to be reinstated as a 'great mind' who was actually close to dropping sexual selection in favour of Morgan's – and Ryan and Jethá's – ideas on primitive human promiscuity.

To this end, this is how they order their quotes from Darwin:

Firstly, they write (p. 42) that Darwin believed "promiscuous intercourse in a state of nature [to be] extremely improbable."

Then (pp. 43-44), influenced, they say, by Morgan, Darwin said that: "It seems certain that the habit of marriage has been gradually developed, and that almost promiscuous intercourse was once extremely common throughout the world."

This (p. 44) is followed by their statement that Darwin *agreed* there were "present day tribes" where "all the men and women in the tribe are husbands and wives to each other."

And lastly (p. 44) a quote from Darwin that: "Those who have most closely studied the subject, and whose judgement is worth much more than mine, believe that communal marriage was the original and universal form throughout the world... The indirect evidence in favour of this belief is extremely strong..."

It sounds, doesn't it, like Darwin was swayed from thinking that Morgan's 'primitive promiscuity' ideas were extremely improbable to coming close to being in agreement with him.

But to quote Darwin properly and in the correct order, he wrote (the third and then the fourth quotes above come first):

"Now it is asserted [note that Darwin is *not* agreeing] that there exist at the present day tribes which practise what Sir J. Lubbock by courtesy calls communal marriages; that is, all the men and women in the tribe are husbands and wives to one another...it seems to me that more evidence is requisite, before we fully admit that their intercourse is in any case promiscuous. Nevertheless all those who have most closely studied the subject...and whose judgment is worth much more than mine, believe that communal marriage (this expression being variously guarded) was the original and universal form throughout the world, including therein the intermarriage of brothers and sisters..."

Darwin then goes on to write:
"The indirect evidence in favour of the belief of the former prevalence of communal marriages is strong, [I cannot find any version that says "extremely" strong as Ryan and Jethá write] *and rests chiefly on the terms of relationship which are employed between the members of the same tribe..."* (my emphasis).

Then Darwin considers the influence of these "terms of relationship" on the ideas about communal marriage:

"The terms of relationship used in different parts of the world may be divided...into two great classes, the classificatory and descriptive, the latter being employed by us. It is the classificatory system which so strongly leads to the belief that communal and other extremely loose forms of marriage were originally universal. But as far as I can see, there is no necessity on this ground for believing in absolutely promiscuous intercourse; and I am glad to find that this is Sir J. Lubbock's view. Men and women, like many of the lower animals, might formerly have entered into strict though temporary unions for each birth, and in this case nearly as much confusion would have arisen in the terms of relationship as in the case of promiscuous intercourse. As far as sexual selection is concerned, all that is required is that choice should be exerted before the parents unite, and it signifies little whether the unions last for life or only for a season."

Now we get to the quote which Ryan and Jethá put second above:
"Although the manner of development of the marriage tie is an obscure subject...it seems probable...that the habit of marriage, in any strict sense of the word, has been gradually developed; and that almost promiscuous or very loose intercourse was once extremely common throughout the world."

Going on to write:

"Nevertheless, from the strength of the feeling of jealousy all through the animal kingdom, as well as from the analogy of the lower animals, more particularly of those which come nearest to man, I cannot believe that absolutely promiscuous intercourse prevailed in times past, shortly before man attained to his present rank in the zoological scale."

And finally his conclusion from which Ryan and Jethá took their first quote:

"We may indeed conclude from what we know of the jealousy of all male quadrupeds, armed, as many of them are, with special weapons for battling with their rivals, that promiscuous intercourse in a state of nature is extremely improbable. The pairing may not last for life, but only for each birth; yet if the males which are the strongest and best able to defend or otherwise assist their females and young, were to select the more attractive females, this would suffice for sexual selection". (Darwin 1871 Chapter XX)

With the quotes in the right order and expanded and corrected it becomes clearer that, while Darwin did give some consideration to the possibility of promiscuous sex, his own understanding of sexual selection and his knowledge of other species made him believe it was extremely improbable.

Most importantly, what Ryan and Jethá omit is the consideration Darwin gives to the classificatory system used by tribes when a father's brothers and male cousins will also be called 'father', leading to the belief by outsiders that any one of these men may be the actual father. We use a descriptive system where 'father' and 'mother' etc. are used for a specific person rather than a class of people, though we might also use these terms in other contexts such as 'father' for priests, godfathers, and 'Father' Christmas etc. without presuming some biological parentage.

Lewis Henry Morgan could not imagine anything other than promiscuous mating preceding the origins of the family, including sexual relations between brothers and sisters which for him would be an inevitable consequence of promiscuous group marriage (something Ryan and Jethá don't acknowledge at all). This he based on what he saw among the Hawaiians who used the same term 'father' to refer to the father, the father's brothers, and the father's cousins. Morgan reasoned that it must mean that all these men in the past were potential fathers, and, together with the use of other such kin terms, he further reasoned that these terms must be survivals of group marriage (Chapais 2008).

But using the same term for members of a class of people does not mean that the term is *descriptive* of an actual or potential relationship in the way we use such a term. When 'mother' is used for a class of females – women of the same generation and residential group – everyone still knows which female is the biological mother. There is no reason to presume there is any greater confusion concerning a biological father (beyond the potential uncertainty of any paternity) when all men of a certain generation and residential group are called 'father'. Morgan assumed the terms must have been descriptive in the past but further evidence did not support this, and Morgan's ideas became understood as misinterpretations of what these kin terms meant.

Returning to the incestuous nature of Morgan's assumptions about our past, when considering our origins in social and breeding groups it has not been unusual to think that incest is normal in animals and was therefore normal in our own past. But this is wrong. Virtually all species avoid incestuous matings (Greenwood 1980). When animals live in social groups, members of one sex (sometimes some of each sex) leave the group at puberty. Usually it is the males that leave but for chimpanzees and bonobos, and, to a lesser degree, gorillas, it is the males who stay and the females who disperse to live and breed in a new group. This male philopatry/female emigration is quite unusual in mammals and a significant piece of evidence for us.

When we think of how sociality itself evolved, the first social unit is mother and offspring. With larger groups we can expect, and most social mammals have, a number of related females with their offspring, grouping for protection against predators and for access to food sources, and males as temporary members of the group. This is what we find in most primates such as baboons, macaques, and langurs, and in other social animals such as lions. And this is what was thought to be our ape origins. For African apes, though, it is the females who leave – groups (communities) of related males stay together for life while natal females leave at puberty and new females come in from other communities.

The dispersal of group members at puberty is the mechanism by which genes flow between populations and therefore throughout the species; without this gene flow populations will be genetically isolated and evolve into different species. It also avoids incest and inbreeding and it may well be the sexual aversion to relatives that impels the individuals to leave. It is something that has most likely been strongly selected because inbreeding often reduces reproductive fitness; individuals that avoided reproducing with close relatives out-reproduced those that did not, and the selection pressure was probably strongest on females who suffer more from a poor quality conception than do males who can immediately mate and reproduce with other females.

The evidence from the other apes strongly suggests that male philopatry and patrilocal residence is at the root of our own evolution. Somewhere along the way we developed the flexibility in where people resided, and we somehow overcame the severe antagonism our male ape cousins express towards males from other communities. There is certainly no evidence to support matrilocal residence at our root. There is, though, all the necessary evidence against any natural incestuous mating in humans which would have made us an extremely odd species.

At the end of their CHAPTER TWO Ryan and Jethá (pp. 44-45) explain what they mean by *promiscuous*. They say that it is not indiscriminate mating or sex with random partners but refers only to having a number of on-going sexual relationships at the same time. In our prehistoric bands of 100 or 150 members at the most, they say that it is unlikely that many of these partners would have been strangers. They say that each person is likely to have known every one of his or her partners deeply and intimately.

We will see later that there are major problems with this picture regarding the existence and consequences of mate preferences, and how this can fit with *shared* sex and the authors' arguments for sperm competition and a bonobo-like human sexuality. But for now we need to think a bit more about these prehistoric bands.

If our earliest groups were like those of chimpanzees and bonobos, numbering fifty to a hundred or so, the adolescent females coming into the group in which they will breed will be starting out there as lone strangers. They will, initially at least, be mating with strangers and living with strangers and not with familiar males as Ryan and Jethá argue. Like Morgan, Ryan and Jethá imagine the group to be a bounded entity through time with members living together and mating together for life. This does not happen in any social species and *if* it did happen in our ancestors an explanation for this unique difference is required. Once we take on board the movement of individuals of other social species in order to mate and breed – something Morgan did not know and Ryan and Jethá have failed to incorporate – we have to include this natural transfer of individuals in our own ancestral scenario.

The idea Ryan and Jethá have that people born in a group of 100-150 people spend their lives together and inbreed with each other has no evidence to support it either in human ancestors or in other social animals. If our ancestors had done this they would have comprised multiple groups in genetic isolation and would have evolved into numerous different subspecies and ultimately different species (and the various different hominin species that *have* existed, though are now extinct, will have come about through such isolation). For a species to remain and evolve as one species there has to be gene flow across the populations within the whole species.

It is important to recognize that a social group is not a fixed, bounded entity in us or in other animals, which is one reason why group selection fails – gene flow across groups within a species mixes the genes and removes the group as a discrete genetic unit on which selection can act.

Perhaps we could argue that, like chimpanzees and bonobos, the immigrant female engages in promiscuous sex with all the males in her new community. But – *unlike* chimpanzees and bonobos – humans evolved extensive kin networks and male-male relations beyond the male's small natal group. This almost certainly depended on the existence of male-female reproductive pair bonds as we will see.

*

- If Darwinian natural selection is to be mocked with regard to humans then we should at least start with some understanding of it as it does apply to other species.

- In sexual selection genes spread when an individual out-reproduces others of the same sex. Two different naturally selected outcomes – the two sexes – then result within the same species.

- Darwin argued for a female role in evolution through active female mate choice. A passive female role, though, is a potential outcome of Ryan and Jethá's argument for casual female promiscuity.

- The sex which invests most in parenting any offspring is a limited reproductive resource competed for by the sex which invests the least.

- Natural selection is a consequence of differential reproductive success.

- Ryan and Jethá distort Darwin's response to the ideas about group marriage, and they fail to explain the impact of the misinterpretation of classificatory kinship terms on Morgan's ideas.

- Ryan and Jethá, and Morgan, miss the movement of sexually mature individuals between groups; they erroneously imagine the group to be a bounded entity through time with members living together and mating together for life.

Many of us may prefer to only look at and think about humans but, like the evolutionary psychology that Ryan and Jethá present, looking just at humans can lead to a disconnection from evolutionary biology and to poor thinking. We need to tackle a bit more evolutionary biology next.

CHAPTER TWO

The Birds and the Bees for Adults

Consider the honeybee in the nuptial flight when a successful male mates with the queen: his 'endophallus' explodes to become a copulatory plug inside her and he drops dead. Why? It is a strategy to prevent other males from mating with the queen but for that privilege he loses both his phallus and his life (Judson 2002). What's more, the queen is able to pop out the copulatory plug and to mate again anyway.

Or look at the solitary bee, *Anthophora plumipes*, where males in their attempts to mate with the foraging females pounce on them, often knocking the females to the ground. This sexual harassment of a female can reach as many as 11 pounces per minute – pounces that rarely result in mating but can reduce a female's foraging success by 50%. Females naturally try to avoid males but the males are waiting for them around the most desirable outer flowers which the females cannot access without suffering the sexual harassment (Arnqvist and Rowe 2005).

The Dawson's bee, *Amegilla dawsoni*, is a burrowing bee. Large males fight ferociously and to the death to mate with the females as they emerge from their burrows. Sometimes females are themselves 'collateral damage' when accidentally killed as they are caught in the crossfire. When all the females have mated the males are all dead, and for the rest of the year life is quite peaceful amongst the female bees. For the male bee it has been a nasty, brutish, and short life but at least, like the honeybee, if successful he has managed to pass on his 'selfish' genes (Walker 2009).

So, what about the birds?

Let's look at waterfowl. The Argentine duck became famous for having the longest penis known for a bird. You can imagine the tittering at this. Most birds do not have a penis but waterfowl do, and what's more its length and elaboration correlates with the frequency of forced copulations. Yes, rape. The longer and fancier the penis is, the more often rape goes on.

Rape can be quite common in waterfowl and it has now been discovered that the females have evolved internal reproductive tracts to counter this harmful male reproductive strategy. The 'penis' is, like those of reptiles, basically a flap of skin which is everted for mating and sperm travels along an external groove. Patricia Brennan has done some fascinating work on the duck reproductive organs and has discovered that while the male's 'penis' spirals in one direction, the female's reproductive tract spirals in the opposite direction. By keeping sperm from reaching its normal depository point close to the eggs well inside, the female can block unwanted fertilizations. When the females accept matings they are able to assist the copulation and help rather than hinder the transit of the sperm.

In some duck species rape can make up about one third of all matings yet result in only about 3% of offspring, so female counter-strategies seem to be effective. These conflicts over matings have produced an 'arms race' where a better defence by the female leads to further evolution of the phallus in the male to try to override that defence. Some duck organs extend 40 centimetres (Brennan, Clark, and Prum 2010).

Gang rape sometimes causes the female's drowning; one study of mallard ducks showed that 7-10% of female mortality was due to male attempts to mate with females (Barash and Lipton 2001, Arnqvist and Rowe 2005).

Female mortality due to drowning during mating attempts has also been observed in other species such as dung flies, frogs, sea turtles, and otters. Male mallards are not immune to rape by other males either; there has even been one observed mating of a male that had flown into a window and died while being pursued by an amorous male who continued with his goal post-mortem (Moeliker 2001).

Male mallards guard their pair-bonded female from forced copulation attempts by other males though at the same time they will attempt to force copulations on other females. Pair bonds usually break up during the first week of incubation as the male, having fertilized the female's eggs, then associates increasingly with other males. This means that many males are again looking for 'free' females and as many as twenty males may pursue a female over long distances in gang rape attempts. The number of eggs in a female's first clutch which are fertilized by her mate will depend on his success in protecting her. If a female loses her first clutch of eggs she will likely re-nest but the second clutch will normally be fertilized by forced copulation unless the female establishes a temporary pair bond with a new male or re-establishes the bond with her old mate (Williams 1983).

Female ducks, like other birds, can store sperm (mallards for 17 days, see Cheng, Burns, and McKinney 1983). A phallus that can be used in forced matings to place sperm close to the egg, just as it is ready to be fertilized, is an adaptation that can 'win' against other sperm, including sperm which the female has stored from her pair-bonded male.

As the song goes: "birds do it, bees do it", but it is not surprising that the song does not reveal *how* they do it.

There is plenty to be seen in nature that we would consider painful or cruel. Darwin noted, for example, the digger wasp which lays her eggs in caterpillars she has first paralysed so the offspring can feed on the living body of the caterpillar. We cannot miss seeing what is involved in predator/prey interactions, or those of parasites and hosts, where natural pain and conflicts of interest clearly exist. Perhaps it is easier to accept the conflict and pain between different species than to understand why they exist within the same species, and especially with regard to sex.

We tend to easily fall for the idea that natural behaviour and happiness are somehow inter-linked, as if doing what comes naturally can only be a good thing. Once we take a closer look at the natural world, though, we cannot help but see that along with the undeniable wonder and beauty there is pain there too. Individuals inflict suffering doing what their genes 'want' them to do but, perhaps more importantly, we tend to overlook how the current bodily home *of the genes that will benefit* is not exempt from this suffering, including with regard to sex. Remember the antechinus![8]

Ryan and Jethá (p. 47) say with regard to humans that there are costs involved in denying one's evolved sexual nature, as if unleashing the casual promiscuity they imagine to be our nature is the solution to all our unhappiness. They quote E. O. Wilson (1978) as if he is in agreement when he says these costs are paid in "the less tangible currency of human happiness that must be spent to circumvent our natural predispositions". But they omit to say that Wilson is writing about a natural human sexual pair bond and certainly not a natural sexual promiscuity. His arguments, which Ryan and Jethá have taken out of context, are directed at the Catholic Church and their attitudes towards birth control, masturbation, and homosexuality, i.e., non-procreative sex.

~~~~~~~~~~~~~

---

[8] If you do need to be reminded, see page 23.

Ryan and Jethá's arguments about evolved sexual traits are limited in the most part to what they think they see in humans, chimpanzees, and particularly bonobos. They give us no indication of an appreciation of evolution or what really goes on in the natural world, and they ultimately only have the current sexy media image of the bonobo as a model for our ancestors.

We'll look to our ape cousins in the next chapter but in this chapter we need to consider the evolution of sex more broadly to make sure that we have a better foundational understanding of evolved, naturally selected sexual behaviours.

Firstly we need to go right back to the very beginnings of sex and the sexes and to follow up something that arose with the Bateman quote in the last chapter. The first part of that quote was:

"The primary feature of sexual selection is to be sure the fusion of gametes irrespective of their relative size, but the specialization into large immobile gamete and small mobile gametes produced in great excess (the primary sex difference), was a very early evolutionary step. One would therefore expect to find in all but a few very primitive organisms, and those in which monogamy combined with a sex ratio of unity eliminated all intra-sexual selection, that males would show greater intra-sexual selection than females." (Bateman 1948)

The evolution of sex and the sexes all starts with *anisogamy*: the evolution of two distinct forms of sex cell which is, as Bateman says, the primary sex difference. Even if we wish to dismiss the relevance of this in our own sexual behaviour we at least need to understand the consequences of the evolution of sex and the sexes over the last six hundred million years.

What follows may seem like a lot to take in and it may seem disconnected from where humans are today but it is still very much about our own direct ancestors and our own evolutionary history. It also gives us an opportunity again to get some insight into the workings of natural and sexual selection and therefore what we might expect to find in the natural world.

Six hundred million years in a few pages

The first life was that of single cells similar to the bacteria cells that exist today. Then a new type of single-celled organism evolved which had its DNA packaged in a nucleus. Both these types of single-celled organisms reproduced by dividing into two, and until about 600 million years ago there were only asexually reproducing single-celled organisms.

These cells are *haploid* which means that they only have one set of genes. Almost all multi-celled organisms today have cells that are *diploid* meaning they have two sets of genes (one set from each parent). When multi-celled organisms produce their sex cells (eggs or sperm) they are producing haploid cells from their diploid cells in their gonads which are the testes and ovaries.

It is also interesting to note at this point that it is only with the production of sex cells that we have death due to ageing. When reproduction is by the division of the parent cell into two daughter cells there is then no parent to age and die because the parent itself has divided to become the offspring. But when we have bodies where only certain of the cells are designated to produce the haploid sex cells for sexual reproduction, the parent body ultimately degenerates over time.

The body is a collection of cells which all have the same genes but build all the different tissue types of the body. Only those copies of the body's genes in the sex cells produced in the gonads have the potential to get into new offspring and give the body's genes their immortality; all other body cells enable the body to function but their genes have no future beyond that body. Senescence, i.e., growing old, arrived only with the evolution of sexual reproduction so death due to ageing is actually the price paid for sex (Clark 1996, Goodenough 1998).

The first haploid sex cells that fused with each other were of the same size. Around the time multi-celled organisms were evolving, the sex cells were also evolving into two distinct sizes. How and why this happened is still debated. One theory is that the limited resources that can be put into sex cells meant that individuals that put fewer resources into each of a greater number of sex cells could, because of that greater number of sex cells, out-reproduce others. As long as other organisms continued to put enough resources in their sex cells for the subsequent development of the zygote (the new diploid cell that results from the fusion of the two haploid cells) then all was well. But as the production of smaller sex cells spread due to its advantage in numbers, these smaller cells were then in danger of fusing with another small cell and the zygote would be at a great disadvantage due to its greatly reduced resources.

There would then be a selection pressure for large cells which would now have the advantage of being a rich resource with a greater certainty of successful development into a new individual. It would ultimately pay to either produce an increased number of small cells *or* to produce fewer but larger sex cells. One or the other size had an advantage – either in numbers or in resources – but an in-between size was at a disadvantage on both counts (Maynard-Smith 1978, Maynard-Smith and Szathmary 1999, Parker, Baker, and Smith 1972).

What we see here is the smaller sex cell – the sperm, the male – reducing investment per offspring and increasing the potential number of offspring that could be produced in total. And we have the larger sex cell – the egg, the female – being virtually forced into investing more in each offspring and reducing the number of offspring that could be produced in total.

This, of course, looks like the way men and women are viewed: the male spreads his investment more thinly across potentially a greater number of offspring, and the female is left to do the bulk of the investment in each offspring, relatively burdened by the reproductive work. It is because of this interpretation that this theory is criticized as being the result of modern views on the relative reproductive roles of men and women being reflected back on the very origins of the two sex cells, and potentially used as evidence for its justification. Males do, however, nearly always produce vast numbers of individually cheap sperm and females do produce far fewer and individually relatively expensive eggs.

It is interesting that this theory of the very beginnings of 'female' and 'male' presents the 'male', the smaller sex cell, as the first to evolve, as if cleverly taking the advantage and virtually forcing the evolution of the 'female', the larger and high-investing sex cell. The male is portrayed as the exploitative sex from the start. With some males fertilizing (and so getting their genes into) many, many eggs while contributing little else to the zygote, males are even viewed as parasitizing the investment in the eggs that is provided by females (Randerson and Hurst 2001).

But there is another theory for the evolution of anisogamy.

The theory for the evolution of anisogamy that I think has more support is one which concerns conflict between the *organelles* of the two fusing sex cells. Around the nucleus of the cell is the cytoplasm which contains organelles such as mitochondria and chloroplasts (in plants) which are essential for the cell's metabolism. These have their own DNA and reproduce asexually within the cell, so when the two sex cells fuse there is the problem of conflict between two genetically different asexually reproducing lines within the same cell which is detrimental to the workings of the cell.

There is good evidence to support the existence of mechanisms that prevent this conflict by allowing only the organelles from one of the two parents to be in the fused cell. There are some primitive species where the organelles from the two sex cells are initially both in the fused cell and battle it out, and where it is the organelles that started out in the greatest number that prevail. In this organelle theory it is the mitochondria that packed more of themselves into the sex cell, and so out-numbered those in the other sex cell, that 'win'. Producing a large sex cell is simply the means to that end (Matt Ridley 1994).

Because this initial battle within the cell harms the functioning of the cell, and therefore also harms the prospects of the genes in the nucleus which come from both parents, there was then selection on the nuclear genes to prevent the battle by letting only one parent cell provide the organelles. Some evidence to support this comes from the discovery that when a mouse or a human egg is fertilized a few mitochondria do enter with the sperm but these have been marked out for destruction by genes in the nucleus *of the sperm itself*. The mitochondria have been coated with a protein signal that leads to them soon being engulfed and digested in the now fertilized egg (Mark Ridley 2000).

It may well have been a combination of selection pressures that led to the evolution of two distinct types of sex cell but the bigger question for us concerns how this also came to mean two distinct sexes. It seems most likely that when the two gamete sizes first evolved it was, in fact, still within only one type of body: a body that produced both types of sex cell so a body that was actually hermaphrodite. Many simpler organisms today produce both types of sex cell so it is quite likely that this hermaphroditism was the ancestral condition. This also means there were no 'males' exploiting 'females', only, if that is how we want to see it, small cells exploiting large cells but with individuals themselves producing both eggs and sperm and therefore being both exploited and exploiter.

Once two types of sex cell evolved it was then because each of them benefited from different traits in their parent body that those bodily traits would be under different, and perhaps opposing, selection pressures. Ultimately, being in a body that specialized in producing only one type of sex cell meant that selection was free to act differently and produce two different body types, each better able to serve the reproductive interests of its respective sex cells.

Being only of one sex meant being a body freed from the constraints of selection for traits that benefited both types of sex cell. Various *sex determination* mechanisms (genetic, chromosomal, hormonal, and environmental) evolved across species so that either a female or a male developed from essentially the same genome (Majerus 2003). Shared genes could then be limited in their expression to one sex or the other and so the same genome could be expressed differently depending upon the sex of the body it is in. The sexed body has been selected to express traits which better serve the interests of its particular sex cell type.

Everything about the two sexes follows from anisogamy.

As well as the simultaneous hermaphrodites that we have already briefly looked at in the previous chapter, there are others that are sequential hermaphrodites. In the sequential hermaphrodites the individuals change in size and other traits when they change sex. The only vertebrate sequential hermaphrodites are a few species of fish. In species such as the monogamous clownfish (of *Finding Nemo* animated movie fame) that live amongst anemones, males are the smaller sex because females can produce more eggs when large. If the female dies the male will grow and become a female and a nearby juvenile male will become the new male in the pair (Francis 2004).

In species where the male can monopolize fertilizations with a number of females the size advantage goes to the male so in these species they start as females and if the male dies the largest female then becomes the new male for the breeding group. As a twist on this, in the blue-headed wrasse there are also some individuals that are born male and can successfully mate as small males. These small males mimic females and are therefore able to swim amongst the females and fertilize eggs even though a large male is present as the 'official' breeding male of the group.

In the invertebrate sequential hermaphrodites the size advantage goes to the female for egg production, for example, oysters first spawn as males and then as they grow they spawn as females. The slipper 'limpet' (not a true limpet) *Crepidula fornicata,* starts out as a male but if he cannot find females he will settle and attach to a rock and become a female himself. Males are attracted and stack on top and as the stack grows the lower males become females. The males higher in the stack fertilize, via long intromittent organs, the eggs of the females lower down in the stack (Judson 2002).

There are also species where the sex of an individual depends on its contact with another individual of its species after an initial sexless larval stage. An example of this is also an example of one of the most extreme size differences between males and females: the green spoon worm *Bonellia viridis*. If the larva has not come across a female after about three weeks it then settles in a rock crevice and grows to a two metre long female. If the larva does meet with a female it is ingested by her and lives in a small chamber in her reproductive tract called an androecium – literally "small man room" – where it produces sperm to fertilize her eggs.

These males never grow bigger than one or two millimetres making them one to two thousand times smaller than the female. As it takes the female about two years to mature there is the risk of dying during that time or even a risk of not finding a mate once matured, so being a tiny male within the female at least avoids those risks (Judson 2002).

There has recently been the discovery of even greater sexual size dimorphism in a marine worm *Osedax* off the coast of California which feeds on the carcasses of whales by digesting the whalebone fats. These worms are all female, about 7cm in size, and are attached to the whalebone. Inside these females are dwarf males of only a fraction of a millimetre in size. Which sex the larva becomes is determined by how many females are already attached – if the bone is already covered with females the larva is ingested by a female and the larval males inside the female's body produce sperm for her eggs (Rouse, Goffredi, and Vrijenhoek 2004).

These species and the sequential hermaphrodites show examples of when being male or being female is linked to a difference in size: sometimes it is better to be bigger as a female and sometimes it is better to be bigger as a male. The green spoon worm and the whalebone-eating worm are examples where being an extremely small male has advantages while the female is much larger and is the only one of the two sexes that can be said to have matured into a full adult. In these species being a tiny male that does not have to spend time growing but can simply settle inside a female and do nothing more than produce sperm is the reproductive advantage that has been selected. For most species the size difference is not so extreme.

Sexually selected traits

Darwin made a distinction between *primary* and *secondary* sexual traits. Primary sexual traits were seen as those that are directly connected to reproduction such as the gonads and copulatory organs, and only the secondary sexual traits such as ornaments and weapons were seen as being the result of sexual selection. But now this distinction is usually dropped and the primary reproductive organs are also viewed as evolving due to sexual selection because the signs are there (sexual competition, mate choice, and sexual conflict) that sexual selection has acted during their evolution.

For example, male copulatory organs (where they exist) vary enormously even between closely related species. This variation is the result of genetic mutations that arise and produce an advantage over the alternatives in getting sperm to eggs which is as much the result of male-male competition (and sometimes female choice) as is the evolution of the other male appendages that gain a reproductive advantage (Arnqvist 1997, Hosken and Stockley 2004).

'Hypodermic' or 'traumatic' insemination occurs in invertebrates such as bedbugs. Though the female bedbug still possesses a normal reproductive tract the males bypass this by injecting sperm directly into the female's body. This causes injury to the female and she has evolved modifications of the abdomen which function to localize the site of insemination to a region which has become packed with immune cells (House and Lewis 2007).

Hypodermic insemination has also been discovered in the rather aptly named *Harpactea sadistica* spider. It seems that this male trait evolved because it bypasses the female's reproductive tract and sperm storage organs and therefore prevents sperm being removed by a subsequent male when the female remates. This reproductive trait, once it arose, will have had an advantage and spread down through generations, replacing the previous 'normal' method of accessing eggs (Řezáč 2009).

Rather than copulatory organs evolving simply to enable successful copulation and being for the benefit of both sexes, it has now become clear that they evolve under sexual selection: these traits are also subject to competition between individuals of the same sex, to mate choice, and to conflict between the sexes over the control over fertilization.

A number of fish have internal fertilization achieved via modifications of the male's anal fin into a groove or a tube-like structure. Selection for this internal fertilization gives an advantage over releasing gametes into the water but it can also be seen as the result of competition between males to get to the eggs first. In shark mating, which appears quite vicious, the male bites the female to hold onto her and females have evolved thick skin as protection against this injury.

Some fish females that deposit their eggs in specific locations also have modified tube-like structures as their ovipositors. In seahorses it is the females that propel eggs into the male's pouch so her ovipositor is therefore an intromittent organ or penis. In the male seahorse's pouch the eggs are fertilized and embed themselves in the walls of the pouch which, like a placenta, has a rich supply of blood vessels to provide oxygen and nutrients. The eggs hatch in the pouch and continue to be nourished there by secretions, their production stimulated by prolactin which is one of the hormones of mammalian pregnancy. After a few weeks the male goes through a couple of days of contractions and gives birth to the young (Forbes 2005).

The ancestors that first left the water to live on land almost certainly had external fertilization and returned to the water to spawn as frogs do today. Other amphibians such as newts have internal fertilization without an intromittent organ: the male produces a spermatophore (a sealed sperm packet or capsule) which he deposits on the ground and the female may or may not take it up into her reproductive tract (Lombardi 1998).

The evolution of eggs with waterproof shells freed our ancestors from the tie to water for breeding but it also meant that internal fertilization was now essential. Snake and lizard males have a pair of 'hemipenes'; crocodiles and turtles a single intromittent organ. In these species the 'penis' is normally kept within the body and is everted for copulation; it is basically a sheet of tissue which is everted to form a groove or tube-like structure along which the sperm can be propelled into the female. Some have spines or hooks to anchor to the female.

The Lake Eyre dragon lizard male bites the female's neck during mating and the male can be so aggressive that he fatally pierces the female's spine. Females try to avoid mating once their eggs have been fertilized but males will forcibly mate them if they can. The female's defence is first to threaten males or try to escape but as a last resort a female will flip on to her back so the male cannot mount her. The female has a bright orange belly and throat, its colour due to testosterone which, along with sustained progesterone levels, is believed to drive the female's courtship rejection behaviour (Olsson 1995, Jessop, Chan, and Stuart-Fox 2009).

In garter snakes many males attempt to mate with a single female as she leaves the overwintering den. Pulsating muscular contractions in the males crush the female so she cannot obtain enough oxygen and her stress response includes a gaping of the cloaca which therefore permits intromission (Shine, Langkilde, and Mason 2003).

In monotremes (egg-laying mammals) the penis is similar to that of the turtle except that the seminal groove forms a proper tube but it is still used only for transporting ejaculate and not waste products for which they still use the cloaca.

Mammals have a penis with a urethra which carries both semen and urine. The mammalian penis varies a lot across different species both in shape and in adornments such as bumps, spines, ridges, and flanges. Erection is either by vascular engorgement as in humans or the penis is fibroelastic and maintained in a semi-rigid state, springing into the female and then back inside the male's body as in deer, sheep, antelope, and cattle (Sparks 1999).

An intromittent organ is not necessary for internal fertilization as we saw with newts and the taking up of spermatophores by the female, or as we see in birds and their 'cloacal kiss'. In spiders the frontal leg-like appendages known as pedipalps are adapted in the male to suck up sperm from his genital opening which is then ready for transfer across to the female.

The genes in both sexes need the other sex in order to have a future but these further examples of mating behaviours show that 'natural' sexual behaviours are not necessarily 'enjoyed' by the two sexes. There can be conflict over if or when to mate, and parental investment beyond the investment in the sex cells can be particularly influential in sexually dimorphic behaviours.

Families

Reproduction is not just about sex but about offspring survival. In a number of species parental behaviour by one or both parents has evolved.

In the penduline tit (*Remiz pendulinus*) either the male or the female will desert the nest, hoping to leave the other parent to raise the chicks while the deserter starts a second clutch. Usually it is the male who deserts first and the female is left to do all the parenting but in 5-20% of cases the male is left as the single parent. Sometimes both parents desert the nest and then the chicks starve: in populations across Europe about one third of egg clutches die due to abandonment. Female penduline tits try to hide their egg-laying from the male in an attempt to lay a full clutch before the male knows how many eggs there are. If he sees that she has laid a clutch then he will desert first but by hiding the eggs the female has a chance to deceive him and be the first to desert (Sparks 1999, Arnqvist and Rowe 2005).

In the context of families we can also note the existence of siblicide. Some birds such as the black eagle, white pelican, and masked brown booby lay two eggs and the parents do nothing while the first chick pecks its sibling to death. In the bald eagle the siblings will engage in fierce fights when food is short but when food is plentiful an older nestling may even tear prey apart and feed some to its younger sibling.

In blue-footed boobies there is also this link between hunger and sibling aggression: only when the alpha chick's growth rate falls below a certain threshold will it kill the younger chick. Similar facultative siblicide occurs in hyenas. In sand tiger sharks the siblicide happens while still in the uterus as it does in the pronghorn antelope where one embryo develops a necrotic tip on its tail and skewers the other embryo behind it (Mock, Drummond, and Stinson 1990, Forbes 2005).

Infanticide by males

While evidence of siblicide is only now gaining attention, evidence of infanticide is well established. Siblicide can be understood, to some extent, as over-production of offspring to allow for failed embryos or variation in resources. Parental infanticide or abandonment can also be seen as parents actively culling offspring according to resources or, as in male fish eating some of the eggs they are brooding, catering for the resource needs of the parent while parenting. But there is also infanticide by unrelated individuals which is not about the availability of food but the availability of reproductive resources, i.e., fertile mates.

In lions, for example, there are usually a few related females with a couple of males in a pride. These pride males keep other males away, which is important because if they are ousted by other males and have young offspring still in the pride the new males will kill those offspring. The pride males are protecting their own genes while the new males are seeking to spread theirs. It is only by killing the young cubs that the females will become fertile again and can conceive the new male's offspring.

Imagine a new male lion in a pride that did not kill the young: by the time the current offspring are weaned and the females become fertile again he may already have been ousted by another male. If not he could soon be, and if the new male killed his offspring he has lost again. The competition is between the different genes via the different traits they produce. A trait for indifference to the lion cubs of other males loses out to a trait for killing those cubs.

Gorilla males also kill the young of other males so that they can then reproduce with the otherwise non-fertile female. It is not any conscious decision making going on here, only traits that have evolved and been selected or not.

When Ryan and Jethá (p. 54) question whether there can be "a discrete genetic basis for something as amorphous as preoccupation with paternity" they would do well to think of these behaviours in the males of so many species that *are* shaped by a naturally selected and completely unconscious genetic "preoccupation with paternity".

This kind of infanticide is widespread across species, including insects, birds and mammals (see, for example, Hausfater and Hrdy (eds.) 1984, van Schaik and Janson 2000). It also occurs in sex-role reversed species: in jacanas the female will kill the chicks and eggs of males when she has taken over another female's territory, and will then mate with those males. In this species it is the male parenting capacity which is the limited and limiting resource.

One extreme example of infanticide due to competition over food resources rather than reproductive resources is the Barrow's golden eye duck which breeds in Iceland. The females have to lead their newly-hatched ducklings upstream to the lake feeding areas and if they pass a male with a female still sitting on eggs the male will attack and kill the ducklings. If they do get to the lake, females who are already there with their own ducklings will also attack the new arrivals. Each year there is carnage with hundreds of dead ducklings (Sparks 1999).

Traits that 'win' in evolution are not necessarily nice. For species where it takes the female (occasionally the male) some time to raise offspring while the males are hanging around largely waiting for mating opportunities, a trait that leads to infanticide plus the ability to subsequently guard one or more females and their young for as long as possible has an advantage. If an infanticidal trait arises in such a population it is likely to spread. Just as genes for strength or speed pass from parent to offspring so do genes for not killing offspring attached to a female with whom you have mated.

Any argument that a concern about paternity must be a recent thing in humans because knowledge of a connection between sex and offspring is relatively recent and requires human intelligence, needs to be set against these kinds of behaviours across species which look very much like a concern about paternity. It would be very hard to explain a lot of the reproductive behaviour seen in males, behaviours such as infanticide, mate-guarding, and male parental care, if they depended on a conscious knowledge of the connection between sex and offspring. Genes are quite capable of being selected without any need for conscious awareness by the individuals they are in of what is going on.

In some species of insects, amphibians, and fish it is the father who does all the parenting (Shuster and Wade 2003). In others both parents are needed, and fathering can pretty much match mothering as is found in the socially monogamous birds with their intensive offspring production. When the discovery came that social monogamy is often not complete sexual monogamy in birds it seemed to undermine arguments for monogamous pair-bonding in humans – but more about that later.

Mammalian females have evolved gestation and lactation and there is room for very little direct male parental care. Many mammals are solitary so the two sexes will just come together to mate, often during a brief mating season, and then part. Some mammals live in small nuclear family groups but isolated from other similar small family groups such as many of the lemurs of Madagascar, the marmosets and tamarins of South America, and the gibbons and siamangs of Southeast Asia. Some live in larger social groups with a number of females and one or two males such as lions and gorillas, and others live in larger mixed-sex groups such as macaques, baboons, and chimpanzees.

Not many of these mammalian males do much, if anything, about providing young with food. Some do, such as the marmoset and tamarin males who carry the young so that the mother can feed herself more easily and sometimes these males share food with young too (Whitten 1987). Males do, though, often provide protection either from predators or from other males of the same species. Fathering behaviour can be selected in the same way as mothering behaviour if it is towards actual offspring of that father and therefore about spreading copies of *his* genes just as mothering is about spreading copies of hers.

Direct and indirect benefits in female mate choice

Females may make mate choices based on the *direct* benefits from males of protection or food, or for the *indirect* benefits of high quality genes in their offspring. The direct benefits are often easily seen and measured, such as in certain butterflies and other insects where males provide nuptial food gifts or package their sperm with some energy rich resources for the female (Andersson 1994). The bush cricket spermatophore can account for over a quarter of the male's body weight and contains a lot of protein for the female to convert onto egg production. The male sage bush cricket actually lets the female chew on his own fleshy hind wings while mating and she feeds on the haemolymph that seeps from the wounds (Sakaluk *et al.* 2004).

One study found that the female of the seed beetle *Callosobruchus maculatus*, though she is harmed by the spiny male genitalia during copulation, mated with more males if she was thirsty – these males provide water with their ejaculate to keep her satisfied (Edvardsson 2007).

According to Ryan and Jethá the exchange of sex for resources makes 'whores' of women which they seem to think is a rightful insult whereas 'slut' they consider to be a compliment because the woman is just having sex in the 'same' way men do[9].

We won't deal much with humans yet but when it comes to other species this exchange of sex for resources, and therefore females as 'whores', is not uncommon. The energy costs carried by the female in reproduction makes the direct benefits of food provided by the male along with his DNA a very important reproductive advantage. Female reproduction is very much about converting resources into offspring, and the acquisition of resources can have the same priority in female reproductive success as copulation has in the reproductive success of males. There can also be little point in the female acquiring 'good genes' if she is nutritionally deficient and fails reproductively anyway.

In many birds and in primates such as marmosets and tamarins, the shared parenting by the male means the female can produce more offspring for them both. At the other extreme is another primate where the reproductive rate is extremely low yet the female carries parenting alone. This is our fellow ape the orangutan where each offspring can take as long as eight years of dependency on its mother before she can start on the next one. This long-haul of single parenthood means that the female is not fertile and is not interested in mating for most of those years of mothering (as well as the low reproductive rate contributing to the potential extinction of the species).

Because of the size of these apes and their ecology there has not been selection for male parental investment which could have increased their reproductive rate. Rape is commonplace too because females, even fertile females, often resist matings either because the male is low quality, or because it interferes with their crucial foraging, or because of disease risk, including sexually transmitted diseases. Knott (2009) argues that because orangutans are largely solitary and females are only mating at intervals of six to eight years they are then

---

[9] Stated by Christopher Ryan in *About the book* in the postscript of the paperback edition of *Sex at Dawn* (2011).

facing a period of significantly increased disease risk during these sexual contacts, and this risk is minimized by female resistance behaviours.

Other than the protection from the nuisance of sexual harassment that a preferred male consort sometimes provides, the burden of parenting is a heavy one carried solely by the orangutan females. It also means sexually receptive females are rare, and suggests that they would be far less rare if female orangutans gained more *direct* benefits from sex and males than they do.

Why are there still 'loser' males?

So with all this selection going on and, presumably, females mating with only a subset of winning males, why does such variation still exist? Why, Ryan and Jethá ask about humans (p. 53), are there still 'losers' in the gene pool today if ancestors were winning males and choosy females? If they are arguing that sexually competitive men and choosy women should lead to the elimination of male 'losers' they only need look to other species to discover that this is not what happens. In only considering humans they mistakenly think this must be evidence against such competitive and choosy behaviours.

Sexual reproduction, mutations, and changing environments all provide answers but it is important to look at it first from the perspective of genes rather than individuals. If we think of evolution as changes in gene frequencies we can imagine all the different genes that exist in all the different bodies of all the different life forms. Individuals are of different species when sexual reproduction cannot work between the two different body types because, amongst other things, the genes in the sex cells are too different to work together. Winning genes are those that get themselves into more bodies than do alternative genes, but their success also depends on the other genes they are working together with in any particular body.

In sexual reproduction the sex cells are made during a process called *meiosis* when the genes in the diploid cell are shuffled so that each haploid sex cell potentially has a new and unique combination. This sex cell then combines with another sex cell which has also gone through the same process which means that offspring and therefore individuals are almost always unique. The actions of genes in development will be affected by the particular internal environment of the body and the external environment comprising other organisms as well as the physical environment. There is always variation created by meiosis plus some new mutations arising from copying mistakes so the combinations that come together in each individual and the subsequent development will always be different.

Because the genes are in pairs (one from each parent, except for those on the X chromosome which, in mammals, males only receive from the mother) it is also possible for harmful genes (alleles) to be hidden by a good copy. This means that it is only when two of the harmful copies of the same gene come together in the same body (or the 'harmful' gene is on the X chromosome in males) does their presence have harmful consequences such as it does with cystic fibrosis in humans. Reproducing with close relatives can increase the potential for two harmful gene copies coming together. Some of these genes can actually be beneficial when there is just a single copy such as the 'sickle cell' gene which provides protection against malaria when there is one copy but is very damaging when there are two copies.

The point is that if we look at it from the perspective of a particular 'winning' individual we overlook the effects of sexual reproduction and the constant creation of variation, and of something new, due to the shuffling of different combinations of genes through different bodies. What 'wins' are the genes that get more copies of themselves reproduced throughout all the potential bodies.

So, we might ask, why not mate randomly? For males this is indeed an option and can even mean attempting to mate with males (for example bed bugs, Ryne 2009) or even inanimate objects such as the Australian jewel beetle that attempts to mate with beer bottles (Gwynne 2003), and sometimes it can lead to mating attempts with individuals of another species such as toads sometimes seen in amplexus with unfortunate fish.

When a lot of sperm are being produced, and making a mating 'mistake' does not have major consequences because correct matings will also potentially be made, then 'indiscriminate' matings are not strongly selected against. It does not serve the interests of genes in males to spend too much time on making sure that it is the correct mate, so it can be better to have a low threshold for a mating response rather than miss a real mating opportunity. This can also include selection for genes that produce traits in males that can manipulate or overrule female mate choice.

For females it is usually different. Mating carries costs for both sexes such as the energy used in searching for a mate, or the time lost that could have been used for acquiring food, or the transmission of parasites and diseases. While we would expect greater risk behaviour in males because of the potential benefits and the potential zero fertilizations, with a limited number of eggs we would expect the female to benefit from mating just enough to fertilize those eggs. When there is a choice of males we would expect that selection will act on that choice where certain preferences lead to more surviving and reproductively successful offspring. As males evolve 'persistence' traits and traits to circumvent female choice, females evolve 'resistance' traits and traits to counter those more coercive reproductive strategies in males.

Being able to choose a healthier male would be a selection pressure. Males can be attractive simply because they are free of disease because catching an infection from copulation itself will be avoided if possible by the female. When females produce a large clutch of eggs we might expect that to have eggs fertilized by different males would be an advantage to achieve more genetic variation in offspring. If we consider species where females benefit from more than just sperm from the male, such as when they provide nutrients or protection, then selection will also act on those choices.

Males may offer good nest sites or even protection from sexual harassment from other males which hinders the female's foraging. In deer, for example, females that are feeding to build up reserves to support their next pregnancy will benefit from a male who can keep multiple other males away – males who are all seeking matings with those females and would greatly interfere with their grazing and therefore their reproductive success as females.

In pair-bonding species we might expect less variation in male reproductive success but we might also expect a lot of variation in both sexes if there is assortative mating where the best male and the best female pair-up then the next best and so on down through the 'mate-value' range. This does not mean, though, that members of each pair would not gain an advantage by mating with others outside the pair bond: a male may gain extra offspring, and a female may gain better genes or a better genetic variety amongst her offspring, or some other resource these other males may provide.

In polygynous species such as lions, gorillas, some baboons, and langur monkeys, there is a winning breeding male or two but their tenure can be brief so females may be mating monandrously (with one male) at any one point in time but are mating with different males, and therefore polyandrously, over time. Though the male may have a number of females simultaneously as mates he only has a relatively brief period in which he can breed. Females are mating over their whole adult lifetime while the males are reduced to a much shorter reproductive period, sometimes a couple of years or less. There will be some males who are completely excluded from breeding but the variation between males might not be as great as a snapshot picture of the polygynous system paints because the females mate polyandrously *over time* – and sometimes with males outside the group too.

Finally we also need to note that high fitness males can produce low fitness daughters, and high fitness females low fitness sons. This is because of traits that are expressed in both sexes but the optimal expression of the trait is different in the two sexes. The usual example of this in humans is the hip which would have been selected in one direction in females to increase width and in the other direction in males for better locomotion. Selection on the same trait is in different directions in the two sexes until another modifier gene allows sex-specific expression. Until such sex-limited expression is achieved a trait that benefits one of the sexes can have costs when expressed in offspring of the other sex (Rice and Chippindale 2001).

Most of the experiments in this area have been with fruit flies and they suggest that many genes remain under opposing selection pressures in the two sexes and that this *sexual antagonism* or *intralocus sexual conflict* maintains genetic variation for fitness and potentially neutralizes the indirect benefits of sexual selection, i.e., the indirect "good genes" benefits (Bedhomme *et al.* 2008, Chippindale *et al.* 2001, Innocenti and Morrow 2010, Prasad *et al.* 2007).

Some evidence of this is now being discovered in other species such as red deer. Male red deer with relatively high fitness have been found in studies to father, on average, daughters with relatively low fitness. Foerster *et al.* (2007) found that selection favoured males that carried low breeding values for female fitness – genes that have an advantage when expressed in males are also expressed in daughters to the daughters' disadvantage.

What all this is saying is that sexually antagonistic selection – the same genes selected in opposing directions when expressed in both sexes – can lead to a trade-off between the optimal genotypes for male and female fitness. The genes are constrained in one sex by the opposing selection pressures they experience when in the other sex so neither sex is able to evolve very far in the direction of their own optimal fitness. This pulling of shared traits in different directions when in each of the two sexes maintains genetic variation. It also helps to explain why the two sexes are often not polarized in many traits: when traits are not completely sex-limited in expression it means that there is a masculine/feminine continuum across the two sexes rather than a clear distinction between them.

We can see why females can be selected to be choosy, males usually less so, and that a smaller number of males than females manage to reproduce. Males are often more focused on mating effort and females on parenting effort though in some species it is the males that parent and females are the sexually competitive and eager sex.

Sometimes when the burden on females of reproduction is looked at it can appear that the males are getting off lightly but it is the males who usually carry the greater burden of competition for mates. In *The Descent of Man* Darwin wrote:

"The female has to expend much organic matter in the formation of her ova, whereas the male expends much force in fierce contests with his rivals, in wandering about in search of the female, in exerting his voice, pouring out odoriferous secretions, etc.: and this expenditure is generally concentrated within a short period... On the whole expenditure of matter and force by the two sexes is probably nearly equal, though effected in very different ways and at different rates."(Darwin 1871 Chapter VIII)

For human females it is difficult to know just how much mate choice they have had throughout our ancestry. In human groups the movement of females, and no doubt males, between groups as marriage partners would be the concern of group members as much as the pair themselves. Important social network links between people in dispersed groups have been made possible by such marital connections, but more on that later.

Evolutionary theory, including 'selfish gene' theory, *does not* say that people in an extended kin group would be selfishly looking out just for themselves; selection acts on copies of the same gene in different bodies so individuals would be expected to aid others who share their genes. Reciprocal altruism would also be selected. But if resources are limited then sharing with the closest relatives would come first. Surely no one would think it strange for a mother to let even a niece go hungry rather than her own daughter if she had to make that choice, and we should not be surprised if selection acts on a male to show a similar preference for those who share more of his genes.

Like the behaviour of the male lions and gorillas towards their own offspring, *natural selection takes the need for actual knowledge of shared genes and actual knowledge of the workings of sex out of the equation.* If male gorillas and lions, and the males of many other species, don't need to know the connection between sex and offspring in order to evolve and carry out their mate-guarding, paternal, and other competitive reproductive behaviours, we can be sure our ancestors did not need to know either.

Extended sexual receptivity and concealed ovulation

Some species only have limited breeding seasons when both sexes become fertile and mate. Females might even have only a matter of hours when they are fertile such as in ring-tailed lemurs. In the apes and many other primates we have menstrual cycles throughout the year if the female is neither pregnant nor lactating, and within this cycle there is a fertile window of a few days around the time of ovulation.

Many species advertise their fertility visually and/or with scents, and male as well as female interest in sex is generally linked to these fertile periods. In some species, and most of all in humans, female receptivity can occur well beyond this fertile window, though, as Frank Beach (famous for his studies of animal, including human, sexual behaviour) remarked:

"No human female is 'constantly sexually receptive'. (Any male who entertains this illusion must be a very old man with a short memory or a very young man due for bitter disappointment.)" (Jolly 1999).

Some have even half-jokingly suggested that for men who are not bonded to a particular woman the condition might be better termed "continuous nonreceptivity" (Alexander and Noonan 1979).

Perhaps things are changing but for now it is useful to consider some important terminology introduced by Beach to distinguish differences in the female regarding sexual behaviour.

Receptivity is the term we have mostly been using so far and it is the original, most-used one, denoting a female's acceptance of a male for mating. This fits with the old idea that females are reluctant maters and passively accept the winning male's advances. *Attractivity* refers to how sexually attractive a female is and therefore how much sexual interest a female receives. Beach introduced the additional term *proceptivity* to refer to female behaviours that initiate and maintain sexual interactions with males; the distinction between *receptive* and *proceptive* female sexual behaviour is an important one which allows for a distinction between a passive acceptance and a more active solicitation of a male (Dixson 1998).

Why have some females evolved to be sexually attractive, sexually receptive – and especially sometimes sexually proceptive – when there is no chance of conceiving? And why do the males mate when there is no chance of fertilization?

In the case of the males it is a response to signals which evolved as honest signals of female fertility and, as we have seen, the threshold for male sexual arousal is often relatively low. Wasting sperm is usually preferable to many males rather than missing a fertile opportunity altogether.

Considering the potential costs of mating to females, that they should have undergone selection for mating when not fertile needs greater consideration. One suggestion has been that with ovulation hidden, only by guarding a female and mating regularly is fertilization assured. The female benefits from the constant presence of the male because she gets protection and perhaps food provisioning or help with parenting, and this is suggested to have enabled pair-bonded relationships in humans (Fisher 1992). Presumably the female continuing to appear fertile even when she has conceived or is lactating is enough to elicit sexual arousal, mating, and mate-guarding behaviours from her pair-bonded male.

It has been pointed out that the pair-bonded gibbons restrict their mating to the fertile periods so constant female receptivity and constant sex is not necessary for pair-bonding. But gibbons live as pairs more or less isolated from other gibbons, and most pair-bonded species (other than birds) also live isolated from other pairs, so we need to consider what living in a larger social group might mean if there are also strong selection pressures for pair-bonding.

A very different suggestion for the origins of concealed ovulation comes from anthropologist Sarah Hrdy who studied grey langur monkeys (not baboons as Ryan and Jethá state, p. 60). Hrdy saw that when a new male took over from another he then proceeded to kill the infants, and it was these studies that ultimately successfully supported the hypothesis that infanticide by males is a selected male reproductive strategy. This infanticidal behaviour has now been discovered in many more species (van Schaik and Janson 2000).

The female langurs were seen to mate with new males even though the females were not fertile, and this is understood to be a strategy to confuse paternity because if a male has mated with a female he is less likely to kill any offspring she subsequently produces. These langurs are polygynous where one male is the only male of the group but only for about two years, similar to what we saw in lions. For a new male to simply wait for the females with young to become fertile again reduces his reproductive success, whereas killing those young of the previous male means the females become fertile again and can be fertilized with his sperm before the next male comes along and ousts him too. As well as mating with the new male who has taken over the group, the females were also seen to mate with males from outside the group as one of these males could potentially become their new alpha male.

By 'faking' sexual solicitation behaviours normally connected to actual female fertility, females of a number of species mate with multiple males to confuse paternity and reduce the potential loss of offspring caused by infanticidal males.

Across mammal species a year-round association of males and females is found in, at most, 34% of species, but in primates the males remain associated with females in 74% of species. Comparative tests have provided strong support for the idea that the risk of infanticide is ultimately responsible for the evolution of permanent male-female association. So the other side of the infanticide coin is that it has led to males associating with the mothers of their offspring and has led to affiliative male-infant interactions in Old World primates (Paul, Preuschoft, and van Schaik 2000).

In hanuman langurs that are in multimale groups as opposed to the one-male groups, it has been found that half of all the defeated alpha males stay in their group and, while previously showing little interest in their offspring, they now defend them against the male newcomers. Studies of baboons have also shown males protecting their own putative offspring in that these males have 'friendships' with the mothers that end if the infant dies.

Infanticide as a male reproductive strategy is not pretty but it cannot be denied that natural selection has acted to produce these behaviours in males – behaviours that do not need any conscious knowledge of the connection between sex and offspring. Evolution is not 'intelligent' with any intentions or goals or purpose. Mutations and sexual reproduction create variation, and those gene variants that produce traits that happen to be better than the alternatives spread over time.

In species with a slow reproductive rate fertile females are not around very often, and offspring are relatively few in number over the female's lifetime, so losing even one is potentially devastating to her reproductive success. Infanticide by males was likely to have been a selection pressure somewhere in our evolution as it is a gorilla strategy and occurs in chimpanzees as well as many other primates. A female who allies herself with a male can gain protection for herself and her offspring, and the protection from who is then the likely father of that offspring is also a male protecting his own reproductive success.

If a baboon-like polygyny within multimale groups (as is most noticeably found in hamadryas baboons, mandrills, and geladas) was also a factor in our own evolution we could then consider the benefits to a female of not having to share a male with other females, and how evolving a more or less constant sexual receptivity and attractivity could elicit more constant attention and mate-guarding behaviour from that male. If we add to this factors that could have prevented polygynous males acquiring more than one female in many instances (such as successful mate-guarding by both sexes) there is the potential for a combination of multiple factors that hold a pair together to have been more strongly selected – producing greater reproductive success for both sexes – than any alternative behaviours, even in multimale/multifemale systems. We'll come back to this later.

*

- The natural world, along with its wonder and beauty, contains considerable natural pain. Individuals both suffer and inflict suffering doing what their genes 'want' them to do, including – or especially – with regard to sex and reproduction.

- Everything about the two sexes follows from anisogamy.

- Many examples of mating behaviours show that 'natural' sexual behaviours are not necessarily 'enjoyed' by the two sexes. There can often be sexual conflict over if, when, or how often to mate.

- When Ryan and Jethá question whether there can be "a discrete genetic basis for something as amorphous as preoccupation with paternity" they would do well to remember infanticide by males and other male reproductive strategies shaped by a naturally selected and completely unconscious genetic "preoccupation with paternity".

- The exchange of sex for resources – direct benefits – is common across species, making 'whoredom' a beneficial reproductive strategy for females of most species.

- The distinction between *receptive* and *proceptive* female sexual behaviour is an important one, distinguishing between a passive acceptance and a more active solicitation of sex.

- Females of a number of species have evolved 'fake' proceptive sexual behaviours to feign fertility, confuse paternity, and thereby reduce the costs of infanticide by males.

In baboons and langurs in mixed-sex groups with multiple males the affiliations between males and females, and males and offspring, are in species where females stay in their birth group and males come and go. What about multimale/multifemale species such as our closest cousins the chimpanzee and the bonobo where it is the males who stay put?

We'll look more closely at these species next and see if, as Ryan and Jethá argue, these species show our own evolution to be without the sexual competition, conflicting sexual interests, and females trading sex for protection or other resources that we see in so many other species.

CHAPTER THREE

Apes

Chimpanzees and bonobos are our closest living relatives. Their line '*Pan*' split from our '*Homo*' line about six million years ago, the gorilla line having split first around seven to eight million years ago (Chen and Li 2001). When we think about these apes we are probably most inclined to imagine something like a modern gorilla as the common ancestor around eight million years ago and then a modern chimpanzee about six million years ago but, of course, the other apes have also been evolving, so what the common ancestors were like cannot be known for certain.

We also need to note that we have had much closer evolutionary relatives who are now extinct. One of the most recent of those extinctions is that of the Neanderthals, but there are many more species that existed after our split from the chimpanzee, some of them our direct ancestors and some our cousins. So if any argument is made that our mating system must be like that of *Pan* because they are the species closest to us today, that argument is made in the inevitable ignorance of the social and breeding systems of multiple other hominin species that no longer exist.

Even Ryan and Jethá in their introduction (p. 11) note that a gorilla-like system was probably the mating system of our own ancestors until *Homo erectus*, i.e., our mating system changed from polygyny only within the last two million years. Their argument for a change to a promiscuous mating system rests on their assumption that this is the only option in multimale/multifemale social groups.

The *Pan* (chimpanzee and bonobo) lineage probably evolved to become multimale/multifemale soon after the separation from our own lineage if the common ancestor did not already have that social system. Evidence about sexual dimorphism and mating systems in our various hominin ancestors is far from conclusive but does suggest that it would be millions of years before our own sexual size dimorphism reduced to our current level. This suggests a polygynous mating system for a significant stretch of our own ancestry but, as we will see when we come to look at evidence from our bodies, there are also many complicating contributing factors for humans in this respect.

We naturally look to our living ape cousins for clues about our evolution but simplistic comparisons need to be handled with care. Ryan and Jethá give no consideration to evolutionary changes along the different ape lineages over millions of years which means that they give no consideration to the selection pressures our own ancestors faced – selection pressures leading to traits such as our bipedalism and particularly helpless offspring – in what were very different environments from those faced by chimpanzees and bonobos.

Male philopatry

We do have something about chimpanzees, bonobos, and gorillas which is significant: male *philopatry* which means males staying in their birth group. If we look at other social primates such as baboons and other Old World monkeys it is the females who are the ones who remain in their birth group. This is the more usual female philopatry where males move to new groups looking for mates. In the case of the African apes something different happened.

Female philopatry is almost certainly the original social system in social species as it is the most common form and is built on the primary mother and offspring unit with the dispersal of sons but not daughters. This changed in our ape ancestry and is unlikely to have happened with a sudden switch-over in which sex leaves the natal group and which sex stays. How could *male* philopatry have evolved? Did it evolve in chimpanzees only after our split from them?

Male philopatry most likely evolved from a polygynous social and breeding system like that of the gorilla where large males attract a number of unrelated females to them. The gorilla is also a close cousin of ours, sharing a common ancestor possibly less than two million years before our own lineage split from that of the chimpanzee, and so the last common ancestor of all African apes is not irrelevant to our understanding of our own roots.

In gorillas we sometimes see a son staying with his silverback father and helping to defend the unit, such as in mountain gorillas where fewer than half the males emigrate. These are therefore multimale/multifemale groups with a polygynous mating system, though the females will sometimes secretly mate with other males in the group besides the silverback.

The females in a gorilla group are not related females who have remained in their natal group but are mostly unrelated adult females (though related females may sometimes join the same male) that the silverback has attracted to his group. The female gorilla allies herself with a silverback and not with other females, and she will change group either because another male is stronger and more attractive, or her silverback has failed to defend her offspring from an infanticidal male, or because the breeding male is a close relative.

In western gorillas it has been found that the silverbacks of neighbouring groups are often related. This relatedness has made possible more regular and more peaceful group encounters than in other gorillas, and it suggests another way male philopatry could have evolved: the merger of neighbouring breeding units, led by related males, to form polygynous multimale/multifemale groups (Bradley and Doran-Sheehy 2004).

In the gorilla we can see polygyny along with elements of male philopatry, multimale groups, and each female seeking to secure her own individual bond with the silverback. In most other species polygyny involves female philopatry, related females, and bonding between related females. Perhaps for the gorilla ancestor the breeding male's tenure became so extended that his daughters as well as sons were selected (due to inbreeding avoidance) to disperse, or perhaps it was the need to find the best male protector against infanticidal males which led to individual females moving to a number of new and different males during their lifetime as each stronger male came along.

In gorillas the females, being unrelated, are each more bonded to the male than to each other so it is a form of male-female pair-bonding – the female's main adult bond is with the silverback. This even suggests an alternative breeding system scenario for our common ancestor with the gorilla: monogamous pairs. With an initial male-female pair bond some males could have been able to attract more than one female, and because bigger males would be more successful in guarding multiple females this would lead to the increase in sexual size dimorphism we see in modern gorillas. Either way, the African ape ancestor some eight million years ago was most likely one that lived in a one-male unit, either with one female or with a number of females.

We then have one or two million years after the divergence of the gorilla lineage when we continued to share a common ancestor with the chimpanzee and bonobo. The multimale system could have evolved during this time from polygynous breeding units where females were already the dispersers. With more sons and brothers remaining in their natal group or neighbouring one-male units joining together, a multimale/multifemale social system with male philopatry can evolve.

What all this suggests is a mate-guarding, polygynous underpinning to the males' mating behaviour within emergent multimale/multifemale social groups but with the benefits that living with related males brings: life-long allies in defence of the shared territory reducing male-male in-group competition.

Male competition can be intense in female philopatric species where unrelated males enter new groups as adults to breed, and alliances between males, if they occur, are often temporary and weak. In our ancestors we had related males staying together for life in their natal group so there would not have been the selection pressure for the more extreme sexual dimorphism which results from competitive immigrant males attempting to monopolize (if temporarily) fertile females living in matrilocal multimale/multifemale groups.

Our common ancestor with the chimpanzee and the bonobo may have been one with strong polygynous tendencies but with the beginnings of a more promiscuous mating system due to male philopatry and the benefits of male-male bonding for joint territorial defence, so possibly not unlike what we find in chimpanzees today. The new environments and selection pressures our own lineage experienced after the split from the chimpanzee lineage could have shifted our own ancestors sometimes towards and sometimes away from a gorilla-like polygyny at different times as their ecology changed.

Before the chimpanzee it was the baboon that was thought to be the likely model for early humans. The savanna baboons have female philopatry and this was assumed to be an inevitable aspect of early social living, leading to imaginings of matrilocal and matrilineal and even matriarchal early humans. Discovering that the chimpanzee and bonobo are patrilocal, as is the gorilla to some degree (and is certainly not matrilocal), means we no longer assume a matrilocal social system for our early ancestors.

There are also some baboon species, such as geladas and hamadryas baboons, that live in very large herds made up of polygynous breeding units. Hamadryas baboons have unrelated females in the breeding units with some degree of male philopatry in that related males can be near to each other with their separate family units. So these baboon species could also support a model for our ancestral breeding system with their polygynous family units existing within a larger grouping of related males.

Ryan and Jethá state a number of times (e.g., p. 64) that monogamy is not found in any social, group-living primate. Polygyny, though, is. Polygyny is also common across human cultures.

Male philopatry and female dispersal is a trait our hominin lineage likely shares with chimpanzees and bonobos. What else our ancestors share with these two cousins, who are often depicted as being as opposite as 'war' and 'peace', is up for grabs.

The chimpanzee model

The studies by Jane Goodall at Gombe, and especially the discovery of tool use and hunting, led to chimpanzees replacing baboons as the model species for early human evolution. The hunting of monkeys by chimpanzees came as a shock when it was first observed because until then chimpanzees were believed to be peaceful and laid-back vegetarians. Their hunting, though, provided a significant connection to humans. Over time we have gained more knowledge about chimpanzees, including from studies at sites other than Gombe. Some differences between chimpanzee populations have been discovered which, along with what little we now know about bonobos, is all information that needs to be added to the mix.

Meat is the one food chimpanzees will share. Ryan and Jethá (p. 67) *quote* primatologist Craig Stanford describing meat sharing at Gombe as "utterly nepotistic and Machiavellian", adding that Stanford says the chimpanzees at Taï share the meat among every individual in the hunting group, whether friend or foe, close relative or relative stranger. Sweet.

What Stanford (2001) actually wrote was:

"Sharing of meat is highly nepotistic at Gombe; sons who make the kill will share with their mothers and brothers but will snub rival males. They will share preferentially with females who have sexual swellings, and with high-ranking females."

At Taï, Stanford says that the captor shares with the others in the hunting party "whether or not they are allies or relatives".

The shift in the 'quote' and the general presentation of what Stanford wrote gives the impression that something very different occurs between the Taï and the Gombe chimpanzees. This is so that Ryan and Jethá can imply that chimpanzee behaviour can vary radically and, especially, that information from Gombe is unreliable.

Looking into this further we can gain more insight into how hunting is different in these two populations due to their different ecologies.

Christophe Boesch writes that the dense forest habitat of the Taï chimpanzees forces them to hunt cooperatively. At Gombe there is less cooperative hunting because the habitat is not dense forest and monkeys can more easily be hunted by lone individuals. So at Gombe a lone individual makes the kill and then shares it. At Taï it is not possible for an individual to hunt alone and they depend on cooperative hunting. If each of the individuals who are the successful cooperative hunters at Taï did not get their proper share of the meat – whether or not they are allies or relatives – then they would lose the incentive to hunt cooperatively and therefore to hunt at all.

Taï females are also important in the support they give in ensuring that the hunters get their reward of a share of the meat, and the females who are the most active supporters of the hunters also get priority of access to meat. Among the hunters, being the captor gives the best access to meat, with other important strategies such as ambush roles also being specially rewarded. Only by rewarding those who successfully use the most important hunting tactics can hunting itself be sustained (Boesch and Boesch-Achermann 2000).

This, therefore, explains why there is sharing "in the hunting group" at Taï – hunting *has* to be by groups whereas at Gombe individuals are able to capture meat by themselves. It also shows that there is not the equal sharing at Taï but quite a complex and political distribution of meat according to hunting role or support for those in hunting roles. The different ecologies have selected for different hunting behaviours, neither of which are disconnected from self-interest and 'politics'. It is certainly not the simple distinction between selfish chimpanzees at Gombe and unselfish ones at Taï that Ryan and Jethá present to their readers.

And then there is the bonobo

Everyone must now have heard of the bonobo. Ryan and Jethá present the two ape cousins pitted against each other as the 'Hobbesian' chimpanzee and 'Rousseauian' bonobo. Hobbes saw the 'state of nature' as being 'solitary, poor, nasty, brutish, and short', whereas Rousseau blamed man's unfortunate condition on the failings of the state – he was not a proponent of the 'noble savage' but of the potential for good, and he saw a failure of society to bring out that goodness.

Ryan and Jethá (p. 69) think that Rousseau "would have considered bonobos kindred souls if he had known of them". This is very unlikely considering what Rousseau wrote of women and sex in *Emile*:

"...The man should be strong and active; the woman should be weak and passive..."

"If woman is made to please and be in subjection to man, she ought to make herself pleasing in his eyes and not provoke him to anger...she should compel him to discover and use his strength. The surest way of arousing this strength is to make it necessary by resistance... This is the origin of...the boldness of one sex and the timidity of the other, and even of the shame and modesty with which nature has armed the weak for the conquest of the strong..."

"... How can anyone fail to see that when the share of each is so unequal, if the one were not controlled by modesty as the other is controlled by nature, the result would be the destruction of both, and the human race would perish through the very means ordained for its continuance?"

"Women so easily stir a man's senses and fan the ashes of a dying passion, that if philosophy ever succeeded in introducing this custom [of an unleashed female libido] to any unlucky country...the men, tyrannised over by the women, would at last become their victims, and would be dragged to their death without the least chance of escape." (Rousseau 1993 edition)

When Ryan and Jethá (p. 39) write: "if women were as libidinous as men, we're told, society itself would collapse" they could, if they had but known it, have been referring to these very same fears expressed by Rousseau himself! It is clear from the above that Rousseau was saying this very same thing about the danger of the female libido; it is therefore odd, to say the least, that the authors present him as having a very different attitude towards women and female sexuality, and that he would have considered bonobos "kindred souls".

~~~~~~~~~~~~~

We still know very little about the bonobo; initially (and still predominantly) our knowledge of them comes only from studies of captive groups. There have now been some long-term studies of wild bonobos but the human conflicts in the Congo and the low numbers of these apes means we know much less about their natural behaviour in the wild compared to some chimpanzee populations. It took quite a number of years of observation before hunting and many other behaviours were seen in chimpanzee populations, and now that hunting has been seen in bonobos, including the hunting of monkeys, we need to take care in making assumptions about them from what so far is only a snapshot of their behaviour in the wild (Surbeck and Hohmann 2008).

Different populations of chimpanzees have been seen making and/or using tools such as stones for cracking nuts, fashioned twigs for termite fishing, and more recently using sharpened branches to stab bush baby prey in their daytime nest holes in trees (Pruetz and Bertolani 2007). There is no evidence yet of corresponding tool use in natural populations of bonobos.

Chimpanzees have clear male dominance hierarchies and a significant degree of within-group male-male competition which is put on hold when the males join together for their patrols of their joint territory. This ability to switch from conflict to cooperation seems to be an advantage of male philopatry as the territorial defence is by male kin who spend their whole lives together.

On the other hand, it also means that the males have not evolved the mechanisms which enable the males in matrilocal social primates to ally with, or at least tolerate, 'stranger' males. These are essential mechanisms in the matrilocal species because the males have to move between groups to breed. For the male philopatric chimpanzees there is both greater cooperation within the community and the greater antagonism towards other communities.

Female chimpanzees seem to be more isolated than the male, foraging alone with their dependent offspring in relatively small ranges, though they appear to be less isolated in the Taï communities (Boesch and Boesch-Achermann 2000). Females are also often dominated by all adult males and can be treated aggressively by males, including the use by males of aggressive sexual coercion.

Bonobo females have higher status facilitated by the alliances between the females even though they, like the chimpanzee females, are all immigrants. The male bonobos are less bonded with each other which has also helped to improve female status, and the mother-son bond can be the male's most important relationship. The male can derive his status from his mother, and mothers can even assist a son in his mating opportunities. This may be less so, though, in the Lomako bonobos where strong mother-son bonds have not been observed (White 1996a).

Some of the chimpanzee violence has been blamed on the provisioning of bananas in order to enable easier habituation, but other non-provisioned groups have now been observed in long-term studies at Kibale, Bossou, and the Taï forest with observations of male violence very similar to those from provisioned communities (Pusey 2001, Boesch and Boesch-Achermann 2000, Boesch et al. 2008).

Bonobos are now 'known' by everyone to be a 'make love not war' species where sex pretty much defines everything about them. At Wamba they are provisioned with sugar cane, and Wamba bonobo parties (subgroups within the community) are larger than those at non-provisioned Lomako so provisioning may well have influenced *their* behaviours, for example, at the feeding sites bonobos show exaggerated sexual behaviour in the way chimpanzees showed exaggerated aggressive behaviour at their provisioning sites.

Bonobos also do have aggressive behaviours ranging from physical attacks such as biting, hitting, kicking, slapping, grabbing, dragging, brushing aside, pinning down, and shoving aside to glaring, bluff charging, charging, and chasing (Kano 1992).

As Barbara J. King (2004) writes: "The media may delight in oversimplified summaries such as 'chimpanzees make war, bonobos make love,' but primatologists know better."

While there are exaggerations of the differences between our two ape cousins there still remain significant differences between these two species, so let's look a bit closer at our forest-living cousins.

We have already noted the highly significant factor of male philopatry and female dispersal to a new group for breeding. Goodall noted that female chimpanzees joined new communities while they were sporting their sexual swelling as this was their passport into the new group – otherwise they can be attacked severely and even lethally by the males (Pusey 2001). Female chimpanzees therefore tend to avoid the periphery of the group where strange males may be encountered. When they do join a new community they often experience considerable aggression from the resident females who are now faced with more competition for community resources.

There was a DNA study at Taï which suggested that over a five year period half of the offspring had not been fathered by males of that community but it turned out that the DNA results were faulty and resident males *had* fathered the females' offspring. The original 1997 letter to *Nature* concerning the DNA study was retracted in November 2001 (Gagneux, Woodruff, and Boesch 1997, Gagneux, Woodruff, and Boesch 2001, Vigilant *et al.* 2001).

In a report in *The New York Times* Gagneux and Boesch said that they concurred that their initial analysis was seriously flawed, and if extragroup paternity occurs Dr. Gagneux said: "it's probably pretty rare". A study of Gombe paternity by Dr Anne Pusey was also reported and their results were that all 14 offspring they sampled were fathered by males within the group (Angier 2001).

Unfortunately, the belief that female chimpanzees are regularly mating with males from other communities had already taken hold. Ryan and Jethá initially repeated this error (p. 70, 2010 hardback edition) but it was removed from the paperback edition after I notified them of the mistake. But in the same context Ryan and Jethá continue to misrepresent work by Anne Pusey (2001) on the sexual behaviour of female chimpanzees. They (p. 69) quote these two sentences from her description of female chimpanzees: "Each, after mating within her natal community, visited the other community while sexually receptive... They eagerly approached and mated with males from the new community."

This quote remains in the paperback edition of *Sex at Dawn* so it is important to reveal what Anne Pusey actually wrote.

Pusey is writing about her studies at Gombe in the early 1970s. By 1970 the habituated community had split into two subgroups which ultimately became two separate communities: the Kasekela and the Kahama. The Kasekela group soon went on to wipe out the Kahama group over a period of a few years.

During Pusey's study *one* female from each subgroup reached adolescence – these are the *two* females referred to as "each" in the quote used by Ryan and Jethá. Both were "fearful of and reluctant to mate with some of the males of their own community". The Kahama female eventually joined the Kasekela community permanently while the Kasekela female returned, pregnant, after six months.

Pusey was studying female dispersal at adolescence. At Gombe not all females disperse which may be due to lack of attractive alternative communities due to human encroachment. At Mahale, Taï, and Kibale she writes that over 90% of females emigrate permanently during adolescence. Establishing that females usually, and males never, disperse has been important in chimpanzee studies.

So Pusey is writing about female dispersal and inbreeding avoidance and is referring only to *two adolescent* females belonging to two subgroups of a recently single community. One subgroup (interestingly probably the stronger group where both females end up) eventually goes on to wipe out the other in a series of vicious, lethal attacks.

Ryan and Jethá take a couple of sentences from Pusey to use as 'evidence' for their story, giving readers an impression of multiple females consistently engaging in promiscuous sex across community boundaries, and therefore giving the impression of relaxed – even friendly – relations between communities of chimpanzees. Presumably they read Pusey's chapter in full before picking out the two sentences to present to the reader so they must know that this was actually about just two adolescent females in the process of transferring, and that it was soon to be the scene of the most vicious inter-community lethal aggression that, as Pusey writes: "permanently changed our view of chimpanzees".

We can also add, again from the Pusey chapter referenced by Ryan and Jethá, that male chimpanzees at the Mahale site killed several infants within their community. These were first- or second-born infants of newly immigrant females from a neighbouring group where all but one of the males had 'disappeared'. The community males had mated with these immigrant females but the females had continued to spend a lot of their time at, or beyond, the edge of the community and away from the community males. Their infants, therefore, were of dubious paternity. Gradually the females spent more time with the males and subsequent infants were not killed.

For chimpanzees the evidence is that females, unless dispersing to a new community with their sexual swelling passport, stay away from the periphery of the community. Females of higher status are, as would be expected, the ones with ranges nearer to the centre of the community's territory. The males, while often competing within the group for dominance, patrol their joint territory together and are very aggressive towards males, and even females, from neighbouring communities. This aggression is sometimes lethal.

For bonobos it appears to be quite different. While the males are also philopatric and the females are immigrants, the males do not form the territorial patrols like chimpanzees do, and the females do not stay away from the periphery of the community. Female bonobos spend more time with other females and in mixed-sex parties, most likely because there is enough food to support larger foraging parties. The females also form alliances with each other which can counter aggression towards them by males, and it means they can gain priority of access to food. Chimpanzee females with dependent young would, if they travelled in mixed-sex parties, be faced with the problem of males reaching the food source first and then monopolizing it, and so they more often forage alone with their dependent offspring in a small range within that of the whole community.

The modern bonobo is confined to a single area, Cuvette Centrale, in the centre of the Congo Basin in the Democratic Republic of the Congo. The Congo River is to the north and the Sankuru and Kasai rivers are in the south. North of the River Congo are the 3-4 subspecies of chimpanzees across central Africa, along with the various subspecies of gorillas. This strongly suggests that the common ancestor of chimpanzee, bonobo, and human (and gorilla) was somewhere in this region north of the River Congo. About one million years ago, i.e., 4 to 5 million years after the *Pan/Homo* split, when hominin species (our ancestors) had long moved far beyond this region, the bonobo ancestor then became isolated from other chimpanzees.

From comparisons of Y chromosome genes involved in the sperm production of chimpanzees and bonobos it has been found that there is much less variation in the bonobo genes. This suggests that a small population of chimpanzee ancestors became isolated in this region south of the River Congo and this small founder population evolved into the bonobo. The low variation in the bonobo Y chromosome genes also suggests some reduction in sperm competition, perhaps due to bonobo females with their relatively higher status being able to exert greater mate choice. Gorillas and orangutans also have low variation in these genes and lower sperm competition (Schaller *et al.* 2010).

There are also seven DNA regions involved in sperm production which in human and gorilla Y chromosomes are identical and therefore evolved in the chimpanzee lineage after the split from ours. This suggests there were strong 'selective sweeps' in the chimpanzee lineage, i.e., strong positive selection for these genes in chimpanzees in connection to their promiscuous mating system, *after* our lineage split from theirs (Perry *et al.* 2007).

North of the River Congo sometime around eight million years ago is where we most likely find the common ancestor of the gorilla, chimpanzee, bonobo, and human. The gorilla lineage split off first but it was after the split of the chimpanzee/human lineage that the chimpanzee went through some rapid changes to the sperm production genes. Then, about one million years ago, a small number of those ancestral chimpanzees became isolated from the others, and the particular genome of this small founder population in this new and separated environment evolved into the bonobo.

For the early human (hominin) ancestor we may well have to give greater consideration to what the common ancestor with the gorilla was like rather than restricting ourselves to the chimpanzee/bonobo dichotomy. We also need to remember that from the time of our split we have had our own distinct evolutionary journey to follow, as has each of the other ape species. The fossil hominins we have discovered that might help us understand this journey are few and far between, and how their clues are interpreted is not without controversy. But we do know that we did come to live in multimale/multifemale social groups, and from this the question arises as to whether we also then evolved a similar promiscuous mating system to the ones that evolved in chimpanzees and bonobos.

Ryan and Jethá (p. 72) write that "crucially, humans and bonobos, but not chimps, appear to share a specific predilection for peaceful coexistence". They say that this is due to a "shared *repetitive microsatellite* (at gene AVPR1A) important to the release of oxytocin" (italics in original).

They then go on to say how oxytocin is sometimes called "nature's ecstasy" and that it is important in feelings such as compassion, trust, generosity, love, "and yes, eroticism". So let's take a look at this 'evidence'.

The paper this refers to (Hammock and Young 2005) looked at the variation in a repetitive microsatellite (repetitive DNA sequence) of the prairie vole vasopressin 1a receptor gene (AVPR1A) and how this related to differences in social behaviour traits of males. Vasopressin is a hormone which is involved in kidney function but has also been discovered to be involved in behaviours such as protection of offspring and male social bonding. Oxytocin, on the other hand, is a hormone mainly involved in female reproduction and lactation and mother-offspring bonding but also in other social bonding behaviours, and it is released, for example, during 'cuddling' and during sex.

The vasopressin receptor gene sequences of chimpanzee, bonobo, and human were also compared in the paper and a region was seen to be shared by humans and bonobos but deleted in the chimpanzee. This suggested that there may be a connection between humans and bonobos and a difference from chimpanzees – but note, this is for a gene for *vasopressin receptors*.

Subsequently, when more primate species were looked at, this region shared by human and bonobo *was also found in the gorilla and in some chimpanzees*. It now appears that it originated in earlier primates, was duplicated in the ape ancestor, and the duplicated region is in all of the apes. The original region remains in the bonobo, human, gorilla, and Central African chimpanzees. It has gone through some different changes in the orangutan and the gibbon, and has been lost in West African chimpanzees.

There is, therefore, no connection only between humans and bonobos. Further, whilst variation in vasopressin does seem to be linked to male social behaviours no relationship has yet been found between these variations and different male social and mating behaviours across primate species (Donaldson *et al.* 2008, Rosso *et al.* 2008).

So, contrary to what Ryan and Jethá write, there turns out to be no "crucial" genetic evidence here about a particular similarity between humans and bonobos regarding oxytocin, a peaceful nature, eroticism, or anything else, and their treatment of all this is (and I'm being generous) shoddy.

~~~~~~~~~~~~~~

In bonobos we have a status of females which appears to be quite different from that of female chimpanzees. How this arose is fascinating to consider because these females are immigrants into communities of natal males just the same as are chimpanzee females. The predominant sexual behaviour of bonobos is g-g rubbing where the females rub their sexual swellings against each other. When a new female enters a community it is not the support of the males she mainly seeks but an alliance with a high status female (Idani 1991). This is achieved via the g-g rubbing, and through this tension-relieving mechanism the new female gains acceptance into her new community.

A study by Hohmann and Fruth (2000) showed that there are rank-related asymmetries in the initiation and performance of g-g rubbing in that it is low ranking females who solicit this behaviour more often, and high ranking females are more likely to be mounter than mountee. This behaviour occurs mostly in the context of feeding and especially when food is limited and contested. It appears to be related to the stress experienced by lower ranking females trying to access food held by females of higher rank: the solicitation of g-g rubbing enables the lower ranking females to get closer to higher ranking females and therefore closer to the food.

Other points to note are that these solicitations are sometimes rejected and the rejected solicitor can become frustrated, sometimes directing aggression towards another individual. Some genital contacts also appear to be enforced by dominants.

Chimpanzees use pant grunts as vocal signals of submission, and in conflict situations female chimpanzees embrace each other. The dominant female may sometimes insert a finger into the vagina of the subordinate or touch her genitals which suggests some mirroring of the bonobo genital contacts. The frequent use of g-g rubbing by bonobos is a mechanism for dealing with tensions due to differences in social status. Social tensions are a more frequent issue for bonobo females because they associate regularly, unlike the much more isolated chimpanzee females (Hohmann and Fruth 2000).

In chimpanzees when male offspring reach adolescence they work their way through the females, aggressively asserting their dominance over each one of them before they then take on any male. In bonobos the males can stay bonded to their mother, and a male's success and status in the group, including his sexual access to fertile females, can depend largely on his mother's status and her influence in the community.

One study showed the bonobo male dominance hierarchy correlating with mating success but that when a mother was present with a lower ranking son he achieved more matings than in her absence. This is because the mothers interfered in the matings of other males and aided their sons when there was interference by others in their matings. Males were also able to get closer to oestrous females by being in close proximity to their mother. In the absence of their mother the lower ranking males were more peripheral, perhaps avoiding aggression from other males (Surbeck, Mundry, and Hohmann 2010).

At Lomako, where strong mother-son bonds have not been observed, Frances White (1996a) states that one male can dominate a small party and exclude other males but as party size increases other males cannot be excluded and individual male-female relationships become more important. At Wamba, Kano writes that in subparties the effect of male dominance is conspicuous and the top male is responsible for 46-77% of matings. He also states that adolescent males with mothers are able to outrank motherless prime males (Kano 1996).

The differences between the two ape species looks to have arisen at least in part because bonobo females are able to travel and feed together rather than being isolated as is often the case in the female chimpanzee. Gorilla females do not form alliances with other females and seem to be largely in competition with them, each seeking to ally herself with the silverback, probably because in gorillas the protection of the silverback is essential to the survival of offspring.

So how did the bonobo come to this arrangement? The particularly extended sexual swellings may well have initially been selected for the g-g rubbing as a counter to the aggression between resident and immigrant females due to competition over resources. Females transfer to a new group at adolescence and at this age they do not have regular cycles but maintain sexual swellings and are almost always sexually receptive (Kano 1989, Kano 1996).

Both g-g rubbing and copulations with males facilitates the young female's entry into her breeding community. The female alliances can then work as an advantageous counter to male aggression in competition over access to food and over matings, and therefore enhance female reproductive fitness. Frans de Waal suggests this may have included the prevention of infanticide by males which happens sometimes in chimpanzee communities but has not been observed in bonobo communities.

Parties of bonobo females can travel and feed together, and with mutual support they can have priority of access to the food sources. Chimpanzee females are more separated and isolated throughout the community territory which is then border-patrolled as a whole by all the males. Bonobo males are often in the foraging parties with the females – the females are attractive to the males because they sport the sexual swellings for much more of the long interbirth interval than do chimpanzees. Along with the g-g rubbing between the females there is also the use of genital contacts and copulations between females and males to relieve tensions within these foraging parties.

Male-male bonding for joint territorial patrols does not exist in bonobos because the males are more often divided across mixed-sex parties with the sexually attractive females. With the mothers also (at least at Wamba) supporting their sons' sexual access to the fertile females the males gain from this mixed-sex travelling too, and they avoid the more severe male-male aggression within the community that is experienced by chimpanzees. This also means that when parties from neighbouring bonobo communities meet they will meet as mixed-sex parties rather than patrolling males meeting other patrolling males – or an individual that can be easily picked off – so inter-community aggression is greatly reduced in bonobos compared to the chimpanzee.

The bonobo pattern of female association and ranging across the territory has meant that there is not the same advantage to bonobo males of cooperative defence of the whole community territory and all the females within it as there is for chimpanzee males where the females are in their own, separate ranges. It has, though, removed the need for bonobo male cooperation for territorial defence and so it has increased the advantages of individual competitive mating strategies for the bonobo male (White 1996b).

When two bonobo communities come into contact females are present on both sides. As females move to new groups to breed it may even be that there is some recognition of individuals in other communities by the females though this has not been documented. The males, who are always strangers to each other, often do display aggression towards each other, and fighting with the infliction of serious bite wounds has been observed at Wamba (Raffaele 2010).

At Lomako high rates of intercommunity aggression have also been seen, and the number of males in a party has been seen to increase on the day following an encounter. In three cases where all-female parties with infants encountered mixed-sex parties from another community the resident females showed fear, screaming and then hiding. They then escaped by running away on the ground and staying silent, and they remained highly vigilant for the rest of the day (Hohmann and Fruth 2002).

So the 'peaceful' inter-community relations of bonobos should not be exaggerated. It is possible that the occasions where less animosity and more 'peacefulness' has been observed is due to the two communities being familiar because they have only recently separated into two new communities, for example. Though lethal aggression has not been observed, non-lethal aggression is common and it would certainly be wrong to paint bonobos as having friendly inter-community relations.

The apparently 'hypersexual' bonobo has greatly influenced our thinking about our ancestors and ourselves but there has certainly been an exaggeration of bonobo sexuality which needs to be addressed, as does the actual promiscuity of both the bonobo and the chimpanzee.

Some aspects of bonobo sexuality do seem closer to that of humans such as the much extended female sexual attractivity and receptivity to mating well beyond the female's fertile window when ovulation actually occurs and conception is possible.

Frans de Waal is perhaps the name most associated with promoting the bonobo as evidence for a more egalitarian and sexy human nature, based on his studies of captive bonobos and especially a group of ten in San Diego zoo. He still argues, though, for the evolution of the human pair bond and the human nuclear family as crucial factors in our own evolution of cooperation which has gone way beyond that of any other species (de Waal 2005).

In his book *Bonobo*, Frans de Waal (1997) writes: "[The bonobo's] sexiness should not be exaggerated. Bonobos do not, in fact, engage in sex all the time. At the zoo, the average bonobo initiates sex once every one and a half hours, whereas the average chimpanzee does so once every seven hours. In the wild the frequencies are no doubt lower. Many of the contacts, particularly those with the very young, are not carried through to the point of sexual climax. The partners merely pet and fondle each other. Even the average copulation between adults is quick by human standards: 13 seconds at the San Diego zoo, and 15 seconds at Wamba. Instead of an endless orgy, we see a social life peppered by brief moments of sexual activity."

In an article written for *eSkeptic* in 2007 de Waal wrote: "I understand the frustration of field workers with the image of bonobos as angels of peace, which is not only one-dimensional, but incorrect. On the other hand, anyone who objects to the occasional hyperbole (such as "chimpanzees are from Mars, bonobos are from Venus"), should realize that no one would ever have heard of the species – and no reporter would have considered them for a piece in The New Yorker – if they'd been described as merely affectionate. Possibly, one or two decades from now a new image of the bonobo will emerge, one more complex than what we have today."

De Waal also notes the distinction between sex and "sex": in bonobos a lot of the sexual interactions are not sex but "sex", that is, brief touching of the genitals or brief copulatory-type behaviour rather than intromission or ejaculation. This is an important distinction which seems to have unfortunately been lost on many who, having heard the words 'sex', 'orgy', and the like, assume that bonobo "sex" is like human sex, including orgasm and male ejaculation. Human sexual behaviour can be without orgasm and penetration too but when we say we have had sex few would think that it had merely involved a few seconds of genital touching without any attempt to reach a sexual climax for at least one of those involved.

The most common form of bonobo "sex" is the g-g rubbing of the females which is where "sex" and sex could perhaps be said to overlap the most as the females do at least sometimes appear to reach orgasm. De Waal says that the San Diego zoo males were never observed to achieve intromission or to ejaculate during contacts with partners other than mature females. Genital massage was only between adult and adolescent males, and oral sex and mouth kissing was in the even younger juveniles (de Waal 1989). Nor did masturbation lead to ejaculation (de Waal 1995). Nowhere have I come across any record of anal sex which, due to the supposed 'anything goes and in any combination' misrepresentation of bonobos seems to be what at least some people may even have been led to imagine is also part of the bonobo sexual repertoire.

Interestingly, the sex-play behaviour of immature bonobos is very much like that of immature chimpanzees (Hashimoto and Furuichi 1996). De Waal also has noted that sociosexual behaviour rates are similar between chimpanzee and bonobo immatures, and that adult chimpanzees use embracing and 'platonic' kissing rather than continuing with the use of "sex" into adulthood (de Waal 1995).

De Waal writes that in captivity low ranking females may copulate throughout the cycle in contrast to wild and high ranking captive females who are most receptive during the maximal swelling period; this again suggests that female rank influences their sexual receptivity, with a lower rank correlating with greater female sexual receptivity. When adolescent females transfer to their new group they are, as already noted, more sexually receptive to males as well as soliciting g-g rubbing from senior females, so de Waal's observations add to this evidence for the use of sex and "sex" by lower ranking females to counter the negative effects of their low status. Sexual receptivity by lower ranking females means they can avoid aggression and gain better access to food.

There is no evidence that "sex" is sought as something just to be indulged in for itself. Takayoshi Kano writing about the Wamba bonobos says that pseudo-copulatory behaviours may be classified as sexual in that they involve contact with the sex organs but these behaviours occur during feeding periods when distances between individuals have to be decreased, rather than during resting periods. While they may appear erotic by human standards they function to ease tensions – and he says that copulatory sex also occurs in the same context (Kano 1989).

So sex and "sex" are connected to the relief of stress during tense situations, usually around food, rather than casually indulged in at any of the many varied occasions throughout the day when sexual arousal and sexual activity might be expected (by humans) to occur simply for pleasure.

Again in *Bonobo* de Waal (1997) writes: "There is a sharp decline in sexual involvement during a male's adolescence due to the tendency of dominant males to occupy the core of traveling parties where the females are. Only when they enter adulthood and rise in rank do males regain access to receptive females. Not that male bonobos are egalitarian with regards to sexual privileges. In contrast to its peaceable image, the species conforms to the general pattern in the animal kingdom of male competition for females... Since the two top-ranking males in any traveling party generally do most of the mating, it is assumed that they suppress the sexual activity of other males."

De Waal goes on to quote Suehisa Kuroda giving a description of feeding at Wamba where young females exchange brief sexual encounters for food: "To me it is always puzzling what males actually gain from these sexual encounters: in most cases the encounters are rather brief and do not seem to end in ejaculation. Nevertheless, it is certain that females 'know' that sex produces enough tolerance in males to allow the females to remove food from their hands. They seem to seek sex for this purpose."

It should also be noted that some of the "sex" involves infants with older individuals, including infant sons rubbing genitals with their mother, so the function of "sex" in bonobos cannot be confused with sex in bonobos or sex in humans which involves sexual gratification rather than being used to ease social tensions. Humans tend to hug and to touch areas other than the genitals as our social tension-release mechanisms and this we share with chimpanzees. All adult bonobo females studied by de Waal at San Diego zoo were full siblings (female emigration not being open to them in a zoo) so the g-g rubbing "sex" here is between full sisters (Hohmann and Fruth 2000).

It is these 'incestuous' and 'pædophilic' aspects of bonobo 'sexuality' along with their very brief and casual nature, and their occurrence during tense situations and not during resting periods, that most clearly mark this "sex" as very different from sex. We would do well not to confuse the two.

So what else are we learning about bonobos in the wild?

Craig Stanford (2000) writes that female bonobos are not more sexual than their chimpanzee counterparts. He argues that there is no difference in frequency of copulation or in the swelling period of the female's cycle when they are receptive to matings and states: "The supposed release from estrus that is said to characterize bonobos has been overstated because the data are based on captive animals."

He also reminds us of how our depictions of primate societies become intertwined with our own political views, as in the 1960s "when the brotherhood of the predominantly male anthropologists foisted 'Man the Hunter' on students and the public alike" to explain the expansion of the human brain, and it was only several years later that female anthropologists "weighed in with the reminder that something had to account for the expansions of women's brains too".

Though captive bonobos do have longer swelling phases than in the wild, studies of bonobos at Wamba have shown that the female's non-swelling phase can still include swellings that are fairly large and attractive to males and that a third of copulation attempts by males were during this phase though half of them were rejected (Furuichi and Hashimoto 2004). Furuichi (1987) reported that 82% of copulations occurred during the period of maximal swelling.

In a comparison of chimpanzees in the Kalinzu Forest with bonobos at Wamba, 96% of copulations were initiated by male bonobos and only 63% by male chimpanzees (Hashimoto and Furuichi 2006). At Gombe it is the male chimpanzees who take the initiative and females are often taken on 'consortships' by a single male; Gombe is also where male sexual aggression against females has been most often recorded. For chimpanzees at Mahale there are no consortships; males are widely dispersed and females often approach males, copulate with them, and then leave, though alpha males are the ones who are most sexually possessive of females.

These results point to a more prolonged attractivity of female bonobos but with little female initiation of sex along with their greater ability to reject males. Chimpanzee females, in contrast, are more often taking the sexual initiative along with being less able to reject males. At least some of the female chimpanzee sexual receptivity is due to male sexual coercion of females and general male intimidation of females (Muller, Kahlenberg, and Wrangham 2009).

One other aspect that is different in bonobos is that females resume anovulatory (infertile) cycles within a year of giving birth – though this has also been noted for some Taï chimpanzees with more than half the females there resuming oestrus within one year of parturition. Taï chimpanzees travel more in mixed-sex parties than do other populations of chimpanzees which suggests a link between females and males travelling and foraging together and selection for extended female attractivity.

This non-fertile sexual attractivity seems to be more the norm for female bonobos. Chimpanzee sexual swellings and matings are concentrated in a relatively short fertile period compared to bonobos, and chimpanzees have a higher rate of copulation at this time than do bonobos. The number of copulations of female chimpanzees and bonobos are about the same overall but those of bonobos are more spread out over the long interbirth interval and, to some extent, over each sexual cycle; this means that bonobo females actually mate less often and with fewer different males when they are actually ovulating and fertile (Furuichi and Hashimoto 2002).

Human females are also sexually attractive well outside their actual fertile window so we could consider that an extended attractiveness in our female ancestors also, as in bonobos, reduced the number of copulations of our female ancestors when they were actually fertile which reduces sperm competition (more on sperm competition later).

When the two chimpanzee sexes are only getting together when the females are fertile, then, as we would expect, females are more sexually proceptive (rather than merely receptive) and matings are more concentrated during these periods (Takahata, Ihobe, and Idani 1996). Female chimpanzees, for their own foraging needs, are normally dispersed and relatively isolated within the community so the need to associate with males when fertile means they then have the costs of increased travel distances and reduced feeding time. They also have the costs of sexual coercion and intensity of male competition for matings in a relatively brief period.

The female bonobo does not have these costs of travel distance and food competition when travelling in mixed-sex parties because of the greater abundance of food sources and the feeding priority she can have over the males. The female bonobo has much lengthier times of sexual attractivity, and ovulation is far less obvious, and she therefore avoids an intense sexual focus on her that a brief spell of sexual attractiveness causes. Sexual aggression or coercion by males would not be a successful strategy when ovulation cannot be detected and therefore the 'reward' of fertilization cannot be connected to that particular male sexual strategy.

Bonobos, probably Taï chimpanzees, and nulliparous (never pregnant) chimpanzee females all show more constant sexual swellings and have what Richard Wrangham calls a 'quiet' periovulatory period (POP) rather than the 'loud' POP of most chimpanzees. He also notes that high ranking Taï chimpanzee females have fewer sexual cycles than do low ranking ones because the high ranking females have less need to use sexual attractivity to gain male support or to appease males; non-fertile sexual attractiveness and female sexual receptivity is a mechanism that counters the potential for male aggression. The evolution of a more uncertain fertile period and increased non-conceptive sexuality seems connected to the costs and benefits of group foraging and female association with males (Wrangham 2002).

We could also again note that increased female sexual receptivity looks to be associated with lower female rank in both species: it is less likely for females to be sexually receptive if they have the social status to be able to reject male sexual advances.

In his book *Bonobo* de Waal (1997) writes: "We have also seen indications from the field that females are serious rivals when it comes to their sons' dominance ranks, and that their fights can be vicious. In other words, bonobo society is not all rosy."

People who have studied bonobos in the wild also have more to add about the full repertoire of bonobo behaviours. Much is sometimes made of face-to-face mating but one study showed only 15% of matings were like this (Kano 1989). Dr Frances White writes that face-to-face mating is not the normal heterosexual adult mating of the Lomako bonobos she has studied but a posture used when a female mates with a smaller, younger male. Most matings in the wild between adults are from behind. At Lomako male rump to rump rubbing is also rare though common at Wamba (White 1992).

Foraging parties at Lomako are smaller than at Wamba (where bonobos are provisioned with sugar cane) and there is less frequent affiliative behaviour. When larger groups do form the males groom for long periods: these are tense interactions rather than relaxed grooming sessions (White, Wood, and Merrill 1998).

White has also seen bonobo aggression: "knock-down, drag-out fights", males especially aggressing against other males, but also males aggressing against females. White says females are subordinate to males, and that with a single male and a single female the female won't win but coalitions of females can win against a male. Females are choosier about mating around ovulation and even form consortships with one male. Mating can also happen just in response to male aggression, so there can be sexual coercion. Foraging males will go ahead of the group and then fight amongst themselves to protect the food patch and females then have to mate with the winning male to get into the tree and feed. Sex is not indiscriminate, nor casual, but is used to solve strategic problems. Most importantly, behaviours vary across bonobo populations as they do across chimpanzee populations and some of these differences relate back to ecology. As White says, ancestral humans will not have been any less variable and affected by ecology.[10]

Another primatologist, Gottfried Hohmann, has also recalled bonobo violence. When a male jumped on a branch near a mother and infant in apparent provocation the female lunged at the male, which fell to the ground. Other females jumped down onto the male in a scene of frenzied violence. After thirty minutes the females all went back up into the tree and the male had gone. During the following year Hohmann and his colleagues tried to find the male, but he was not seen again and is presumed to have suffered fatal injuries.

On another occasion a newborn was taken from its mother by another female and the next day the infant was back with the mother

---

[10] From an interview with Dr Frances White by Eric Michael Johnson on his blog *The Primate Diaries,* http://primatediaries.blogspot.com/2007/07/frances-white-on-bonobo-behavior.html

but was dead. It is presumed that the female who had taken the infant had kept it from being fed (Hohmann and Fruth 2002, Parker 2007). This, incidentally, is behaviour that has been seen in some monkeys where a high-ranking female will take the infant from a low-ranking mother and due to neglect the infant will die (Silk 1980, Maestripieri 1993).

So where does this leave us? Hopefully it is somewhere with a better picture of bonobos than the simplistic 'peace and love' version or the crass imaginings of constant human-like sexual orgies.

It is most likely that the separation of a relatively small group of chimpanzees south of the Congo River led to the evolution of the modern bonobo traits over the last 1-2 million years and their species-specific behaviours such as g-g rubbing. Perhaps that original rainforest founder population already had more mixed-sex parties foraging together, and the immature sociosexual behaviours were somehow carried on into adulthood. Extended sexual swellings would have already been present in adolescent females, their irregular and anovulatory cycles aiding their entry into a new community of strangers, so this extended attractivity could have facilitated not only nonaggressive relations with resident adult males but also with the females. By reducing the tension over food competition between resident and immigrant females, and with larger food patches to be shared, the females were able to travel and forage together and to form alliances that could give them mutual benefits in countering male dominance, aggression, and male food priority.

Males no longer benefited from coming together to jointly patrol a territory which no longer contained widely dispersed and largely solitary females, and they spent more time with females on a daily basis. This meant they could also sustain relationships as adults with their mothers who now had influence which could benefit the male in gaining access to oestrous females when his rank in the male dominance hierarchy was relatively low. With males more individualistic, and mothers remaining important to sons, the females gained even more in relative status and influence. Anovulatory cycles in the long interbirth interval kept females attractive to males in these mixed-sex groups and "sex" as well as sex helped to reduce tensions between females and between the sexes and enabled the females to access the all-important food resources that could be translated into offspring.

If we wish to argue that five or six million years ago our common ancestor with the chimpanzee already had these behaviours we then need to ask why the chimpanzee lost them in the last million years or so. As de Waal says, if the common ancestor showed bonobo-like pansexuality why did it disappear in the chimpanzee and human? (de Waal 1995).

North of the River Congo are the chimpanzees and the gorillas and it is there where our lineage will have split from that of the chimpanzee/bonobo ancestor. We were soon to become bipedal and to make our way out of the dense rainforest – or the forest retracted and left us outside. Changing conditions require innovative measures to survive. We may still have been in gorilla-like groups with one or two males and a few females and immatures. But at some time, if we were not already, we did come to live in larger mixed-sex communities most likely with male philopatry and female emigration to new communities at sexual maturity.

The ape in the mirror

In relation to the use of the bonobo by Ryan and Jethá, what does this extra information tell us?

Bonobo sexuality is not, as they argue (p. 63) "turbocharged" and "utterly divorced from reproduction" – otherwise there would not be the male hierarchies and sexual exclusion of subordinate males, nor mothers aiding their sons in their mating opportunities, nor females being choosier when ovulating. Sex does, though, seem to be strongly linked to access to food for the females, either food held by males or by higher-ranking females. There is not a lot of sex or "sex" going on outside the context of access to food.

Sex, as Frances White noted, is not 'casual' but extremely strategic, and female bonobos turn out to be far more like strategic 'whores' than the pseudo-male 'sluts' portrayed by Ryan and Jethá.

*

- Ryan and Jethá give no consideration to our common ancestor with the gorilla or to evolution along the different ape lineages over millions of years; they only have a single human ancestor model: the modern bonobo.

- Though monogamy is not found in primate social groups, polygyny is.

- Ryan and Jethá present Rousseau as a man who would have considered the bonobos "kindred souls" when Rousseau clearly feared that unleashing the female libido would lead to the collapse of society.

- Ryan and Jethá distort Anne Pusey's work to use as 'evidence' for general chimpanzee female promiscuity across community boundaries, and relaxed, possibly even friendly, relations between communities. This 'evidence', though, concerns two adolescent females in the process of group transfer, their communities also providing the first shocking observations of male inter-community violence when one group eventually wiped out the other in a series of vicious, lethal attacks.

- Genetic evidence supports the evolution of bonobos from a small founder population that became separated from other chimpanzees.

- Genetic evidence supports selection acting strongly in the chimpanzee lineage in connection to promiscuous mating *after* the *Pan/Homo* divergence.

- Ryan and Jethá use a vasopressin receptor gene to falsely create a "crucial" oxytocin link between humans and bonobos; this vasopressin repetitive microsatellite is in gorillas and Central African chimpanzees as well as humans and bonobos.

- The clear distinction between sex and "sex" in bonobos is not made by the authors and a false sense of adult human-like sexual behaviour is presented.

- Bonobo sex is very much in the context of food and female bonobos turn out to be strategic 'whores'.

What is genuinely significant about both chimpanzees and bonobos is their contrast to humans: in our evolution we evolved some mechanism whereby males from different natal groups were able to interact more peaceably and move between groups as did the females.

And somewhere down the line we also evolved in a way that meant that parenting from more than just the mother became essential.

Enter the male-female pair bond?

CHAPTER FOUR

Paternity

Ryan and Jethá start their Part II of *Sex at Dawn* by presenting their evidence for what they say is a lack of interest by men in paternity certainty. This 'evidence' is to support their story of promiscuous social and breeding systems in pre-agricultural modern humans of the last 200,000 years (and, p. 11, a "few million" years before that time too). In their story these ancestors lived without sexual competition, sexual jealousy, or pair-bonding, and resources were equally shared, including that crucial yet undefined resource of 'sex'.

Partible paternity in the Amazon

One of the last places to be reached by modern humans was South America, most likely only within the last 15,000 years during which time small founder populations diversified into numerous different tribal peoples. One of the interesting finds across many of these people is their belief in 'partible paternity' which is a belief that the foetus is built from the accumulation of semen over time, and therefore this can be from one man or many.

Ryan and Jethá begin their CHAPTER SIX with a brief quote explaining this belief, from the book: *Cultures of Multiple Fathers, The Theory and Practice of Partible Paternity in Lowland South America*, edited by anthropologists Stephen Beckerman and Paul Valentine (2002).

Ryan and Jethá then insert their own *unsupported* statement (p. 91) that a mother will "typically seek out sex with an assortment of men. She'll solicit 'contributions' from the best hunters, the best storytellers, the funniest, the kindest, the best-looking, the strongest, and so on – in the hopes her child will literally absorb the essence of each. " They follow this by saying that there are similar understandings of conception and foetal development among many South American societies and provide a list of such societies from Beckerman and Valentine's book.

So let's take a closer look at each of these 'partible paternity' *horticulturalist* tribes (where women are often regarded as providing nothing more than the receptacle in which the foetus grows) as they are described in Beckerman and Valentine (2002). Bear in mind while reading that Ryan and Jethá have led readers to believe that these are examples of tribes where women "typically seek out sex with an assortment of men" soliciting semen contributions "from the best hunters, the best storytellers, the funniest, the kindest, the best-looking, the strongest, and so on", and note the disparity between that statement and what follows.

Cashinahua

The Cashinahua allow for discreet extramarital sex but public acknowledgement is rare – in the 42 years from 1955-97 only seven men were publicly recognized as sharing paternity with a husband. With regard to extramarital sex in general, it is mostly instigated by men, and women rarely find the brief and furtive liaisons enjoyable. For women it is more about feeling desired and acquiring gifts from the men, especially meat for the married women. Men are less tolerant of their wife's sex outside the marriage, and extramarital sex by either spouse can cause tension, with subsequent conflicts even leading to the fissioning of a village.

Girls are usually sexually inexperienced when they marry around puberty, and promiscuity lessens a girl's chance of securing a husband so she risks economic failure. Nursing mothers are prohibited from having sex for up to a year after a birth and this is often a reason the husbands give for seeking sex elsewhere. Women may seek sex outside the marriage to get back at the husband but more often they refuse sex to their husband as his punishment. When a second father is publicly announced it is only after careful consideration of the risks and benefits to those involved, the latter being things such as the extra meat that will come from the other father, and the social and economic networks that can be opened up between all those concerned (Kensinger 2002).

- Women have sex outside marriage for meat, not pleasure.
- Men are less tolerant of a wife's extramarital sex.
- Sex outside marriage can cause tension, conflicts, and even split a village.
- Promiscuous girls are less able to secure a husband.
- Wives use extramarital sex or refuse to have sex with their husband as punishment for his extramarital sex.

Curripaco

The Curripaco believe in partible paternity but for them there are strict rules against sex before or outside marriage, and the foetus grows from the accumulation of semen only from the husband. People cannot subsist without a spouse due to their economic interdependency. Marriages are arranged by older men and neither of the pair have much choice in the matter. Marriages are linked to alliances and can be used to neutralize potential enemies; if symmetrical exchanges are not honoured warfare can result.

Sex cannot be openly talked about and if a woman has sex with various men they say there is the risk that no one would recognize the child. When the child is *everybody's* they mean in effect that it is *nobody's*. Some people in practice are able to manipulate their situations and have more freedom over their sexual behaviour, but if a girl becomes pregnant outside of marriage she will be sent away to relatives to give birth secretly and will likely kill the newborn – even starving it to death. Sometimes brothers may share the same wife but this is unstable because of conflict and jealousy (Valentine 2002).

- Strict rules against sex before or outside marriage.
- When the child is *everybody's* they mean in effect that it is *nobody's*.
- Babies born outside marriage are killed.
- Conflict and jealousy when a wife is shared which sometimes occurs with brothers.

Ese Eja

The Ese Eja comprise three distinct groups with a long history of warfare and intermarriages. In the Ese Eja marriage is a basic social relationship they all seek to attain and maintain: it sanctions sex and establishes responsibility for children as well as organizing economic and cooperative relations. In extramarital sex men are expected to give gifts of meat or fish or trade goods. Spouses can become jealous and irritated by affairs. Until recently marriage was between a man and his sister's daughter but this is now viewed as incest.

The foetus is believed to grow from accumulated semen and if a husband does not consider that he resembles the child he will not feel any responsibility for it and it may be given up for adoption. In one community 14 out of 27 children reputed to have secondary fathers were given up for adoption. Some men insist on sexual exclusivity and men rarely refer to their partible children as their own.

Other than the gifts women get in exchange for the sex there seems to be no further benefit to the women of partible paternity other than the way it weakens and undermines kinship links through men (and therefore male dominance) because relatives often shun the child of a kinsman if he is not considered to have a substantial share in the child. Another consequence in this small population (numbering about 1200) is the increased number of possible blood relatives partible fathers creates. This limits the number of potential mates, and for girls especially there is the risk and fear that she will unknowingly seduce her own secondary father as well as possible siblings (Peluso and Boster 2002).

- Women acquire meat, fish, or trade goods in exchange for sex.
- Spouses become jealous and irritated by affairs.
- If a husband does not consider that he resembles the child he will not feel any responsibility for it and it may be given up for adoption.
- Some men insist on sexual exclusivity.
- Men rarely refer to partible children as their own.
- There is a fear that girls will have sex with their own father or brothers.

Kayapo

In the Kayapo the principle of partible paternity is recognized but it is uncommon in practice. One reason for this is that when a child is ill it is important for the father(s) to undergo food restrictions, so if the fathers are not recognized they will fail to do so and the child (it is believed) will die. 'Partible paternity' here applies more to the sequence of husbands and lovers a mother may have over time – there is both a high divorce rate and a high mortality rate.

There are some ceremonial sexual practices (more about these types of practices in the Canela below) and some punitive use of rape of girls for being too reticent about sex or for gossiping about sex they have had with married men. It used to be common for a girl as young as seven to be married to an adult man and to sleep with him, though without sex until puberty.

Men who have sex with unmarried women or girls without providing them with game are criticized as stealing them. Sex is represented as a service that women provide to men and for which they are entitled to recompense (Lea 2002).

- Partible paternity is rare in practice.
- Some punitive use of rape.
- Sex is viewed as a service that women provide to men in exchange for resources.

Kulina

In the Kulina, as in other small populations, numbers are so few (sometimes as few as 20-50 individuals) within the group that people are divided simply into those who can be sex partners, including spouses, and those who cannot. In the past marriages were arranged between cross-cousins but these are no longer enforced. Everyone is seen as brothers and sisters (a by-product of ancestral kinship systems persisting in small populations) which further complicates relationships in these very small and reproductively isolated groups.

The Kulina have a belief that men need meat much more than do women and children, so many children have chronic protein malnutrition once past weaning, and child mortality rates are about 40%.

These inward-looking (and breeding) groups consider distant groups barely human: either dirty, thieving, sexually loose, or hostile. Some small family groups move away from the larger grouping for political reasons or to avoid witchcraft, though they can then struggle because larger households with more than one man to bring in meat do better than smaller ones.

When a small number of people are so closely related we might expect reduced jealousy but any 'other fathers' still remain discreet, and men do not like to openly acknowledge that their brother may in fact be another partible father of their child (Pollock 2002).

- Chronic protein malnutrition of children.
- Hostility towards outsiders.
- Sexual jealousy and only discreet extramarital sex.

Ye'kwana

Though the Ye'kwana believe, or believed, that accumulated semen built the foetus this is to be from one man only. Sex is an intimate subject not to be talked about, and female fidelity is important – if discovered the wife is likely to be abandoned (Rodriguez and Monterrey 2002).

- Semen from husband only.
- Wife likely to be abandoned if sex outside marriage is discovered.

Piaroa

The Piaroa believe that semen is a substance of power but there is no role for it beyond providing the original seed, so for them there is no possibility of partible paternity. These are a people who value control of the appetites and infidelity is not acceptable. A child with no known or acknowledged father is destined to die through neglect or infanticide (Rodriguez and Monterrey 2002).

- No possibility of partible paternity.
- Infidelity not acceptable.
- Fatherless children die through neglect or infanticide.

Secoya and Siona

The Secoya and Siona recognize only one man as the genitor. Families are quite dispersed so extramarital sexual opportunities are few but women are still rarely left unattended. Sexual 'excess' is seen as risky for men, women, and the foetus, and various sexual taboos influence sexual behaviour.

Parenting is emphasized over sex, and a large majority marry for life. The men think of sexual intercourse as using the woman and that they must respect their wives and not 'use' them too much. Girls are cloistered, chaste at marriage, and shy in sexual matters while womanizers have a low standing in the community (Vickers 2002).

- Only one father is recognized.
- Women are guarded.
- Risks are associated with sexual excess.
- Emphasis is on parenting not sex.
- Sexual intercourse is seen as using women, and respect for wives means they should not be 'used' too much.

Warao

For the Warao the father's semen initiates a pregnancy and nourishes the foetus but there is no sense of a requirement for semen from multiple fathers and it is viewed as having negative effects on the foetus. If a baby does not resemble the father it may be rejected by him. Men are extremely sexually jealous and oppose extramarital sex by wives except in exceptional circumstances, such as if the man is ill or infertile. There is a preadolescent mortality rate of 47-50% (Heinen and Wilbert 2002).

- Semen from men other than the husband is viewed as having negative effects.
- Men may reject babies who do not resemble them.
- Extreme sexual jealousy.

Yanomami

The Yanomami view the foetus as developing over time due to the work of the semen; viewing sex as work means that it is not good to have sex too often. If there is a second father it is usually the brother of the husband, and sometimes men can be jealous and angry and reject wives due to 'other fathers'. When a secondary father is acknowledged he is obligated to help provide for the mother and child.

Seventy per cent of women have only one of their children 'co-fathered', with another 26% having two, so the vast majority of children have only one father. Of the children who have more than one father 88% have no more than one secondary father (Ales 2002).

- Sex is viewed as work and so it is not good to have to have sex too often.
- Jealousy and anger and rejection of wives due to 'other fathers'.

- The vast majority of children have only one father and the vast majority of the rest have only one secondary father.

Bari

In a study of the Bari it was noted (as it has been in the Aché) that a secondary father was associated with a heightened probability that a pregnancy would eventually produce an adult. The main component of this survivorship advantage was found to take place before birth due to lower 'foetal wastage', most likely due to the gifts of fish and game given to the woman in exchange for sex. Women who had lost a child were more likely to take a secondary father for a subsequent child (Beckerman *et al.* 2002).

- Women acquire meat and fish from men in exchange for sex which increases survivorship of pregnancies.

Matis

Erikson writes that in 1986 there were only 109 Matis due to the epidemics of the 1970s which left most people widowed or orphaned. One consequence has been the loss of *any* potential mate (potential mates being cross-cousins); he was told about one man who had sex with dead donkeys and tree sloths before stealing a girl from a different tribe.

The men who 'share' women are again 'brothers' (actual brothers or male paternal cousins). Fathers and sons formerly sometimes shared the same female because little girls would move to live with the prospective father-in-law (the maternal uncle) who was meant to raise the girl until she was old enough to marry the son. Though technically incestuous, the uncle would have sex with the pre-pubertal little girl 'to make her grow' and as part of her 'education'.

Ceremonial extramarital sex is only implied, rather than verified, due to the ban on sex with regular partners at these times. Erikson also notes in his chapter that there are many instances across Amazonian tribes where women who are too promiscuous are believed to risk semen overload and thus the dire consequences of twins or extra big babies, or of sickness and death.

Erikson also writes that men sometimes put a child's life in danger because it is the child of another man, and he witnessed an angry young man throw his baby son in the river after a quarrel with his wife and her lover (the baby was rescued). He says such cases may be widespread and notes the *often dire consequences of paternal jealousy in the region* (Erikson 2002).

- Hardly any potential mates at all.
- Uncles formerly having regular sex with their pre-pubescent nieces.
- Dire consequences faced by women who are too promiscuous.
- Male sexual jealousy and anger and the subsequent risks to children across the region.
- From this chapter Ryan and Jethá (p. 94) tell readers only the juicy titbit that some Matis may be admonished for being 'stingy of one's genitals'.

Canela

For the Canela we have a wealth of information thanks to William and Jean Crocker (1994).

The Canela are a tribe living in a single circular village of about 1000 people. Females stay in their natal home while males move to the home of their wife. Couples are instructed by their uncles not to be sexually jealous. Girls have their first sex between the ages of 11 and 13 and the young male (perhaps ten years older) she has that first sex with automatically becomes her husband. There are many steps she has to go through before her marriage is secure, the final one being the birth of a child.

After a few months of 'marriage' (the couple might not see much of each other) the girl is then expected to accept sexual invitations from other men. After doing so she is then assigned, along with another girl, to a men's society during a summer festival, and is expected to allow perhaps as many as 25 men to have sex with her sequentially. She now wins her maturity belt which, after a brief period of seclusion, is painted by her female in-laws. Her role now is predominantly to provide sex to motivate men to work and bond. Sometimes she receives some meat from them or other small gifts for her female kin.

When the girl becomes pregnant her 'freedom' comes to an end. She now is burdened with work as a mother in the extended maternal household, with only occasional opportunities to engage in extramarital trysts when she has no nursing infant. While the father also has new responsibilities he is still working and hunting (and formerly engaging in warfare) with other men so his extramarital sexual activities with childless girls, including sequential sex, continues.

Girls who become pregnant soon after marriage, that is, soon after their first sex, are greatly pitied because of the burden of motherhood and the restrictions on their lives. When a girl or woman does become pregnant and she has sex during the pregnancy a number of times with other men, these men are also seen as fathers though with fewer obligations than the husband to the child.

Male bonding within this single community is effected through quite severe training of youths plus the bonding through 'sharing' of wives and sequential sex during summer festivals with childless girls assigned to them for the purpose. The young male at puberty goes through food and sex restrictions that may last several years as part of his training. Until it was stopped in 1940 the hazing/shaming ceremony of youths was used to instil respect and fear in young men, including being used as punishment for having sex with girls.

Youths also go through various internments, instruction, and discipline to train them for military-style obedience. Sex is actually the most effective means of social control for the Canela – especially social control of young men. The Canela were once fierce fighters and their socialization and intergenerational authority is based on their need for military discipline.

The Canela extramarital sex system means that the less attractive men can at least have sex on the festival occasions of sequential sex which levels the men to some extent and improves male bonding. This institutionalized sex for male bonding and alliances, perhaps an expansion of the partible paternity belief, seems to have prevailed (mostly) over open expression of sexual jealousy and possessiveness in the Canela in part because they live in a single bounded community of life-long relationships and inter-dependency as well as being militaristic. They do, though, have to be instructed not to be jealous. The loss of warfare led to youths no longer submitting to control by the elders and ultimately the male-male bonding and the suppression of sexual jealousy broke down.

What is particularly telling is the information on female orgasm. Crocker concludes, and I have to agree, that female orgasm does not occur. The majority of the sex is for male gratification, and as masturbation is strictly forbidden for both sexes the fact that most sex lasts for a matter of seconds is perhaps not too difficult to believe. Canela sex involves no kissing and no touching of genitals, and the man simply squats between the legs of the woman who is not to move, his hands to each side of her body for support.

It is perhaps no wonder that girls and women might volunteer for sequential sex as it is only after a succession of *very* brief copulations that they may experience some arousal and pleasure. The women most enjoy sex with husbands or long-term lovers when more time is taken, though the women are still meant to be very restrained.

What is even more surprising is that when questioned, the men also denied experiencing anything like the mild convulsions of orgasm that were described to them. Only a very few men on occasions were said by the women to become 'dizzy' or to 'quiver', and this again was mostly when taking time with a wife or long-term lover.

Sex is discreet, though sexual performances are discussed with others so there is no hiding any unusual, and therefore unacceptable, sexual behaviour such as touching or women moving in obvious pleasure. Care is taken not to embarrass spouses or their kin. Women and girls can refuse sex to some extent and the men do often experience sexual rejection. But girls have to accept much of the sex to gain their belts and achieve full marriage and therefore to have babies.

About every other year there is an occurrence where a girl will not agree to sequential sex and so she is forced to comply by a group of men each having sex with her to 'tame' her (basically a punitive gang rape). She knows she has no choice and that even if injured she will gain no sympathy: she has been raised by her aunts (it is less difficult for aunts than for mothers to instruct the girls on these matters) to know what is expected of her and that she will have to comply. She is also told that she may even come to enjoy it, and especially to enjoy sex with a long-term lover in time.

- 'Sharing wives' needs to be understood in the context of pubescent girls becoming the wife of their first sexual partner and then the wife of each subsequent sexual partner until their first child is born.
- The sexual sharing of childless girls enables male-male bonding in the context of social control of young men and military-style discipline for warfare.
- Masturbation is strictly forbidden for both sexes.
- Extremely brief sex with little or no enjoyment for the women – and possibly not that much more for the men.
- Sexual enjoyment more likely for both sexes in long-term sexual relationships.
- The need to be instructed not to be jealous.
- Discreet sex.

- Punitive gang rape if a girl should refuse to take part in sequential sex when assigned to do so.

Ryan and Jethá (p. 103) only quote one paragraph from *The Canela* where Crocker and Crocker explain the importance to the Canela of sharing, including sharing the body, so that readers might understand why women would choose to please the men, and men to please the women who express strong sexual needs. This selected paragraph follows Crocker and Crocker (1994) writing that a man of elite standing is able to "catch" a number of women for private trysts but a man of poor standing finds it harder so the occasions of sequential sex on workdays give him a sexual outlet and motivate him to work. This equalizes men to some degree and contributes to male bonding.

Crocker and Crocker say that to some extent men are exploiting women but women are drawn into it by the expectations of gaining a belt, a husband, and a family. They say that as masturbation is not allowed the sequential sex occasions are a valued sexual outlet especially for those men who find private trysts difficult to obtain.

Following the paragraph quoted by Ryan and Jethá, Crocker and Crocker write that what they have traced in that chapter, i.e., how the Canela are socialized to bond through sex, may seem to amount to severe coercion but they argue that we are all coerced in our own cultures. In most societies, they say, there are penalties for not conforming and so it is for the Canela: "the stingy person receives few favors from others... The ultimate penalty is extreme. Some rejected shaman throws a spell of illness which may eventually result in death."

Regarding the punitive rape of girls who refuse to comply with sequential sex, William Crocker earlier writes that he prefers not to use such a negative word as "rape" when the girl could have avoided the forced sequential sex experience by obeying the mores of the tribe. She is not forced to have sequential sex to humiliate her, nor is it an expression of hate or disrespect of women, but it is to bring her to accept sequential sex as a female associate to a men's society. Being such an associate is a prerequisite step to further steps into marriage.

Is this the institutionalized sharing of sex for the common good that Ryan and Jethá envisage in our ancestors? I mentioned earlier that their argument is potentially one of passive females and the removal of female mate choice, so is this how they imagine it to be: childless girls used for male-male bonding until they become young mothers?

A significant part of Ryan and Jethá's argument is that sexual rejection of males leads to sexual frustration which makes them angry and competitive so sexual rejection should be greatly reduced if not eliminated. Yet in the Canela this sexual use of girls to assist male bonding in the group has been for the purpose of producing fierce warriors for warfare purposes against other groups, not laid-back guys.

Very chimpanzee-like!

Mehinaku

Gregor's book (1985) on the Mehinaku has the title: *Anxious Pleasures*, which, from what he writes, is apt. There are some similarities with the Canela in being a mainly in-marrying village, in the seclusion of pubescent boys and girls, and in the potential for numerous sexual partners, though husbands and wives are sexually jealous and affairs are fraught with danger.

Like the Matis, the Mehinaku use the term 'stingy with their genitals'; in this case it is specifically used by men about the women as many of the men are constantly sexually frustrated. Men exchange gifts for sex and will even do so with the established lover of a woman, not only with the woman herself.

The sexuality of females is linked to their inferiority and the disgust with their bodies and menstrual blood. Sex is brief with little enjoyment for the women, and orgasm for women is unlikely. In spite of the potential for many sexual partners men (perhaps not surprisingly) often find it difficult to find a willing partner. As Erikson (above) mentioned more generally for these tribes, in the Mehinaku promiscuity for a woman is seen as potentially dangerous and may lead to the birth of twins or some other abnormality, in which case the infant or infants are buried alive.

As Ryan and Jethá (p. 103) noted for their readers, on one visit Gregor estimated that there were 88 extramarital affairs going on amongst the 37 adults, counting each pair as two affairs. *Actual* sexual encounters, though, are modest in frequency, limited by long taboos associated with rituals and the life cycle, by the absence of privacy, by competition from jealous husbands and more attractive rivals, and especially by the difficulty of finding a willing female partner. Several of the young men *say* they are able to have sex on a once-a-day basis but Gregor says "the frequency of sex for the average Mehinaku, however, is far less".

The Mehinaku are one of a number of tribes in this region where the men have sacred flutes and the women are gang-raped, or threatened with gang rape, should they set eyes on these flutes. Their mythology is one of an original matriarchy, the men having seized power from the women. Since then women have been kept in their inferior place with gang rape as one threatened punishment for not accepting male rule.

Gregor writes that the women are fearful of the men and their potential violence, and even have nightmares about it. There is also a thin line between consensual and forced sex as the man often takes the woman by the wrist and she is 'dragged off' for sex, the term for this 'dragging off' being the same term as that for rape by an individual man (there is a different term for gang rape). As Gregor says, "the system is maintained by the threat of phallic aggression".

- Sexual jealousy, and affairs are fraught with danger.
- 'Stingy with their genitals' specifically used by men about the women as many of the men are constantly sexually frustrated.
- Men exchange gifts for sex.

- Female sexuality is linked with female inferiority and disgust with female bodies.
- Little if any sexual pleasure for women.
- Risks to foetus of female promiscuity.
- Low frequency of sex.
- Gang rape a potential punishment for not accepting male dominance.
- Women are fearful of men and their violence.
- The system is maintained by the threat of phallic aggression.

The Aché

Ryan and Jethá (p. 92) give a little more attention to the Aché referring to Hrdy (1999a) where she quotes from Kim Hill the names for different fathers according to their different contributions to the growing foetus. Ryan and Jethá tell the reader about the *gratitude* rather than jealousy and rage the men of all these partible paternity societies are likely to feel towards these other men for "pitching in".

So let's expand a bit on what Hrdy says about the Aché. Hrdy (1999a, pp. 246-248) writes:

"Men who provided the mother with meat while the baby was forming are seen as especially likely to have given the child its essence."

"Among the Aché, Hill observed that children with just one father received less help, but when a mother lined up *too many* fathers, the extreme uncertainty of paternity dissuaded all candidates from helping. Children identified as having one primary and one secondary father had the best survival rates" (emphasis in original).

"The optimal number of fathers under these demographic and subsistence conditions appears to be two. Are husbands jealous? Among the Aché, men deny it, but then later beat their wives. Not surprisingly, Aché mothers try to convince possible fathers that the club is more exclusive than it really is."

So, there does not appear to be much evidence there for that male *gratitude* Ryan and Jethá tell the reader exists!

On page 236 Hrdy writes:

"Among the Aché an infant who lost his father was four times more likely to die before the age of two. Even if the father was still alive, Aché children whose parents divorced were *three times more likely to be killed* than if the marriage endured. When a widowed or abandoned mother takes a new mate, risks to her infants shoot up. Terrible prospects are one reason why some foraging peoples bury orphans alive along with the deceased parent" (emphasis in original).

"Among the Aché mothers themselves sometimes kill fatherless infants after a conscious evaluation of what the future holds."

Bobbi Low (2001) in *Why Sex Matters* also refers to the poor prospects of fatherless children amongst the Aché.

"[W]hen a man dies, his young dependent children are far more likely to die than if he had lived. While reciprocal sharing of meat is ordinary, when a man has died reciprocity can no longer be extended, so meat is no longer shared with the widow and children."

- Jealous husbands beat their wives.
- Fatherless children, including those of divorced parents, are sometimes killed.
- Meat is not shared with women and children who have no husband/father to join in the hunt.
- The optimal number of fathers is two.

Ryan and Jethá have at least got something right when they note that the presence of another male who may be the father is a form of insurance for these children. In the Aché just one other possible father is the optimal number and meat providers are preferred. As we have seen, meat is the main incentive for sex for women across these partible paternity tribes.

In *none* of these partible paternity peoples listed by Ryan and Jethá is paternity shared in a very open or widely distributed way.

Some of them are very small in-breeding populations, and with the close relatedness of the men, sometimes actual brothers, we might expect reduced jealousy but even in this case jealousy does arise. Extramarital sex is meant to be discreet when it occurs and it most often involves men exchanging gifts for sex from women; for the women it is primarily about acquiring those resources, sexual pleasure often all but absent, so again it is more about being 'whores' than 'sluts'. Nowhere does the reality match the picture painted by Ryan and Jethá.

Death from war or disease has been a big part of the lives of many of these people, so some acceptance of partible paternity may well represent a form of insurance against the death of a husband/father. And in spite of the benefits partible paternity can potentially bring, such as this insurance against the death of the husband, or to allow for more males to have more sexual variety, or to level the potential variation between men in their sexual access to young women, or to facilitate male-male bonding, none of the examples present anything close to a solution to the male-female relationship problems arising from sex, marriage, and children. As we have seen, there are plenty of negative consequences too, and in most cases having more than one secondary father is a woman pushing her luck too far and is certainly not increasing her husband's gratitude towards increasing numbers of other men for pitching in!

Partible paternity also has the effect in some of these tribes of making sex *always* about reproduction because the semen is always believed to be contributing to a foetus. For the Yanomami discussed above, there is no disconnection between sex and making children; some of the men in these tribes talk about it as hard work and they call all sex 'making children'.

We can also imagine partible paternity being favoured by men of higher status as a means to allow them a significant degree of polygyny without the lower status men being totally excluded. Partible paternity allows for higher status men, especially the better hunters, to get more sex with more women without disrupting a basically monogamously egalitarian marriage system. Marriage is clearly almost always essential in that the traditional division of labour between husband and wife creates the basic economic unit on which both men and women depend.

Beckerman and Valentine (2002), writing about the tribal variations, say that they can be looked at as outcomes of contests between the sexes over their relative control over reproduction. Women do best when they remain in the households of their mother; men clearly dominate when they have patrilocal residence. The different economic roles of the two sexes and the influence of warfare are important factors too, as is the particular status of individuals and the kin networks they have, including those created through marriage (see also, Walker, Flinn, and Hill 2010).

For the South American peoples with such small tribal numbers, when it comes to marital residence it can be a matter of simply moving across the village circle. Usually marriage for humans means movement to a new location for one of the pair at least. The evidence from our ape cousins suggests that at the very root of our ancestry it was females, and not males, who emigrated to a new community of 'strangers' to then reside and breed. Would it still have been possible for a lone pubescent female to travel some distance to make her choice of breeding residence when our early ancestors moved from the cover of forest into more open ground? It seems unlikely. Perhaps these hominin females changed residence when different groups came within range at watering sites.

In modern foraging groups married people of either sex may move to live for different amounts of time in different groupings within an extended kinship network but we do have a couple of little hints that for much of our past it was most likely the young females who went to live with their new mates:

A study of 19 *Australopithecus africanus* and *Paranthropus robustus* teeth, dating from roughly 2.7 to 1.7 million years ago from two caves in South Africa, found that more than half of the female teeth were from outside the local area while only about 10 per cent of the male hominin teeth were from elsewhere. While it would not be right to conclude too much from such little data it does fit with what we would expect from what we know of the dispersal mechanisms of our ape cousins, and that even at this time in our past most incoming group members are female with far fewer males leaving their natal group (Copeland *et al.* 2011).

A study of the DNA from a group of Neanderthals from a site in Spain showed that each of the three adult females had a different mitochondrial lineage but the three males had the same one. Though the Neanderthal ancestor would have split from our own more than 500,000 years ago this still may well be representative of a patrilocal mating system of that common ancestor and one that has predominated throughout our ancestry (Lalueza-Fox *et al.* 2010).

Apart from the partible paternity peoples in South America there are no others elsewhere with similar beliefs which would have made for interesting comparisons. The one other society mentioned by Beckerman and Valentine is the Lusi in Papua New Guinea who also have a belief in the primacy of the male role in growing babies from accumulated semen, and, at least potentially, it can come from more than one man.

The belief in partible paternity probably goes back to a founder population in South America which may have been very small. Perhaps over time with the spread of numerous distinct small tribes they each depended on strong in-group cohesion to survive in a difficult environment which included significant tribal warfare, another feature of South American societies (Beckerman and Lizarralde 1995).

Tribal warfare likely led to a female biased adult sex ratio (Mesoudi and Laland 2007) but, along with a male primacy belief in the accumulation of semen making the baby, 'shared' paternity rather than more obvious male hierarchies and polygyny prevailed. The in-group cohesion is increased by closely related people in the villages marrying locally or within the village rather than there being a wider movement of members creating alliances between more dispersed groups through inter-marriage. The in-group distinction between one man and another, or one woman and another, could be blurred to the extent of there being little difference between close relatedness due to blood or to marriage.

Yet still in most cases the 'sharing' of women, especially of paternity, is far from easy, and men almost as much as women focus their parenting effort on their own progeny. What stands out too is the importance of marriage for all these peoples and it would be hard to overstate the interdependence of men and women created by the sexual division of labour; this also means that sex, either within marriage or in discreet extra-marital affairs, is traded by women and girls for the resources that only the men have.

~~~~~~~~~~~~~~~

Ryan and Jethá (p. 93) see a similarity between partible paternity and something Desmond Morris described in his book *The Soccer Tribe* where the team-mates are happy to share a girl with whom they have 'scored' (no mention is made of the perspective on this of the 'girl'). This 'sharing', though, is nothing to do with paternity and the sharing of the actual or potential mothers of their children.

In the Amazonian tribes, *everybody's* child is *nobody's* child and will have very poor prospects, even killed by the mother as we have seen; sharing a girl or woman in this way results in the loss of any sense of obligation or responsibility for any resulting child or for any other consequence faced by the woman. It is not stated how many of the soccer team-mates who shared women went on to stay and support children with those women but we can be pretty sure that a guess of zero will be close to the mark. This treatment of 'girlfriends' is simply men 'enjoying' using young women for sex without any long-term interest in, or concern for, these 'girls'. The scoring in this game is: male bonding: 1; paternity: 0.

In small groups, such as we have seen in the Canela, young teenage girls are used on ceremonial occasions for male bonding but once they become mothers, most likely conceiving with their husband (with whom they would have regular sex) and by the end of their teens, that all changes. It is as if the young, childless girls have a communal 'sex-worker' role to satisfy the polygynous inclinations and bonding needs of the men, and when they become mothers they then have the more usual and much less promiscuous 'wife and mother' role. This can be seen as another version of the Madonna/whore division of women, only with the same females having both roles, changing from one to the other with the advent of motherhood. It also promotes a male-bonded 'brotherhood' for work and warfare; the 'gang rapes' (actual or threatened) in some of the tribes are also a way the men assert their dominance and strengthen their male-male alliances.

Ryan and Jethá (pp. 93-94) continue their story with an argument that *Socio-Erotic Exchanges* (S.E.Ex. for short!) were crucial in our ancestral groups for the binding of adults into groups that cared communally for children with obscure or shared paternity. One piece of evidence they give as a remnant of this "shameless libidinous behaviour" is what British sailors and travellers apparently experienced in Tahiti in the 18th century.

They present (p. 95) a couple of sexy quotes about 'shameless libidinous' girls, taken out of context from *South Sea Maidens: Western Fantasy and Sexual Politics in the South Pacific* by Michael Sturma (2002). One of these quotes is from an account of James Cook's voyage by John Hawkesworth, published in 1773, who wrote about a "young man, nearly six feet high" performing sex in public with "a little girl about 11 or 12 years of age". Ryan and Jethá omit to add that Sturma goes on to note that it was later said that this couple "were exceedingly terrified, and by no means able to perform".

Sturma also quotes from subsequent visitors to Tahiti "that the prostitutes are only a particular set among the rest", the women were far less liberal with "the last favour" than previously supposed "though the offered bribe was ever so great", and none were married women.

But, as Sturma writes in his book – which is a book about the creation of sexual fantasies about 'South Sea maidens' – "the myth of Tahiti and 'free love' had taken on a life of its own".

Unfortunately, it is a myth that is gladly perpetuated by some to this day. It can be amusing how supposedly 'free' sexual behaviour by 'dusky maidens' in foreign lands can engender fantasies in the minds of men. Tahiti is one such fantasy. Women, especially the young, unmarried, poorer ones, trading sex for the novel goods that sailors brought to their island (something familiar to seaports around the world) is likely one part of what contributed to this fantasy native sexuality.

Sturma argues that these constructions of "native" feminine behaviours provided an idealized antidote to Western women's self-assertion. More importantly though, he writes:

"From the beginning the mythology of the Tahitian maiden depended not only on publicizing her attributes, but on suppressing or omitting certain aspects of early European contact."

Central to the mystique, he says, is the image of canoes full of bare-breasted young women coming to greet the ships.

"What such accounts generally omit are the coercive elements that underpinned early European contact with Tahiti. When Samuel Wallis...first made landfall at Tahiti in June 1767 they were not greeted by canoes full of willing maidens but attacked [and] numerous islanders were killed by canon and musketry... Only after admitting defeat did the local Tahitian chiefs send women to placate the invading Englishmen. Wallis recorded in his logbook that 'notwithstanding all their civility I doubt not that it was more thro' fear than love that they respected us so much'.

"Indeed, the first offers of Tahitian women were used to gain strategic advantage [by luring the crew ashore]... [T]he firepower displayed by Wallis in this initial encounter with Tahitians does much to explain the apparent warmth and generosity that greeted later visitors."

Indeed it does.

With this in mind it would probably be best not to set too much store by the embellished accounts of the sexual experiences of relatively powerful Westerners with 'a particular set' of young women in foreign lands whether those men are holding weapons, goods, dollars – or notebooks. Today we don't have to look far to find similar attitudes to, and stories about, the women and children in Far-Eastern tourist destinations.

Clearly there is nothing about the nature of the human female that causes her to have some genetically determined trait for mating with only one male in her lifetime. Few other females do – even those female animals that are monogamously paired may mate with an extra-pair male should an attractive (or threatening) one come along *and* the resident male does not see (or cannot get rid of him for her). And that is a clue to what goes on in monogamous pairs: the mate rarely if ever will be *indifferent* to an extra-pair copulation by his or her 'other half'.

Like much else in life, what an individual is able to do in his or her own self-interest is constrained by the self-interest of other involved parties. In humans the husband's self-interest in mating matters has largely held sway over the wife's because of the asymmetry in size, reproductive role, and social status. Male philopatry – males staying with close relatives, females moving to such 'brotherhoods' on marriage – has likely played a role in this too. Once we have arranged marriages, especially as these affected young girls being exchanged between different groups as a means to forge alliances (and often leading to conflict and warfare too) then the role of these marriages is beyond the 'mate choice' of those to be married.

It is hardly surprising that the threat to the social group of adultery then looms large because sexual desire outside of these intergroup marriages becomes significant, especially when that desire is emanating from a wife in an unwanted arranged marriage consummated at puberty (possibly to a much older man) and supposedly for life. In societies today that severely punish adulterous women it has to be kept in mind that the willingness of these wives to commit adultery in the face of such punishment is not a sign that they wish to mate with multiple men, as Ryan and Jethá (p. 98) argue, but only reflects the drive to make *their own* mate choice. This is not evidence for female promiscuity, only for the strength of the female drive to choose her own mate.

The false promise of promiscuity

Female sexual behaviour, in humans as well as in other species, tends to be dependent on the situation rather than being a more constant spontaneous drive as it is in the male with his sperm-serving sexual behaviour. This is hardly surprising in that a body that is producing something in the region of 2,000 DNA packages a second, every second of every day, is prepared for a very different potential sex life than one that is producing a single package a month – and then only when no fertilization has yet occurred, no pregnancy is in place, and no infant is being nursed.

True sexual monogamy is uncommon in both sexes across species but the reasons for seeking multiple mates are different. In our evolution offspring needed more from males then merely not to be harmed by them, so something more than a benign indifference from males has been selected: either direct resource provisioning (food, protection, education, training, social support) of offspring or indirect provisioning through their mothers.

Chimpanzees and bonobos do not have the male parental investment in offspring that is found in humans. They do not have the pair bonds that are found across human societies, nor do they have the sexual division of labour that is universal in humans. They also, as we have seen, do not have friendly relations with males outside of their small natal group. As our closest living relatives they therefore have very different reproductive behaviour and very different social networks.

Ryan and Jethá (p. 99) suggest that paternity *un*certainty could have been selected for as a trait that increased a child's chances of survival. Paternity uncertainty in other species does increase the survival prospects of offspring in that a male, having mated with a particular female, will not kill her offspring. But when it comes to paternal care or provisioning of offspring, paternity uncertainty acts against, not in favour of, the evolution of such parenting traits in males. The *more* uncertain men were of paternity the more they would take an active interest in the child? Hardly. This, as we have seen, is certainly not what happens in the partible paternity societies, while in chimpanzees and bonobos where paternity *is* that uncertain there is no male parental interest at all – it isn't selected for; it can't be.

What we actually find in all those partible paternity societies is the same as we find everywhere else: "the more likely copies of my genes are in you the more likely I care for you". When the members of the group are closely related then yes, there is some degree of positive care in relation to that general relatedness, but there is always a bias towards those most closely related.

Even if we start from an *imaginary* position of shared paternal care of all offspring it would only take genes in one male to gain an increase in reproductive fitness by biasing his resources towards offspring of a mother who biased her mating more towards him and those genes would spread and displace the alternatives. This male paternal bias in provisioning, protection, and support would in turn increase the reproductive benefit to the mother of convincing such a male of his paternity.

For women who can achieve this focus from a male, all well and good. Other women may indeed have to take what they can from a number of different males in a more direct exchange of sex for occasional and more meagre resources. It is quite likely that there would have been a number of different mating strategies for both sexes but what is least likely – nay, impossible – is that communal sex together with communal paternal investment shared equally across all offspring could have existed. Natural selection does not work, and has not worked, that way.

Either Ryan and Jethá's argument is for a chimpanzee/bonobo promiscuity where the benefit to offspring is that males do them no harm and the mothers do everything else, or their argument is for some degree of mate choice and preferential pair bonds which would mean the males have some recognition of their most likely offspring and selection would act to bias paternal benefits towards those offspring as we see in the partible paternity societies. On a sliding scale of potential paternity we are not going to get selection for biased male care towards those offspring a man is least certain are his.

There is a vast difference between not harming an infant because the male has mated with the mother, and positively directing resources towards that infant or child. Hunter-gatherer males today do not harm offspring of women in the group with whom they have not mated so we can posit that our ancestral males did not need to have mated with a female to behave in the same, benign way. Either females had the protection coming from a recognized attachment to the father of their offspring or a male may have made himself attractive to females by showing his ability to provide for and protect offspring.

The display of male parental care in this latter case acts as the male's 'mating effort' rather than 'parenting effort'; the male presents himself as a potential protector if the mother is not paired, or a better one if she is. Males protecting offspring that are not their own can therefore be the way a male attracts a female and gains an opportunity to father her next offspring. This kind of paternal care as 'mating effort' is found, for example, in some baboons and other monkeys (see, for examples, Smuts and Gubernick 1992).

Marlowe also looked at male parental care and mating effort among the Hadza foragers. He found that biological children received more care than stepchildren, and men provided less care to their biological children as their mating opportunities with other females increased. This is therefore a combination of parental care biased towards own offspring, parental care towards stepchildren as mating effort regarding their mother, and the trading of parenting effort for mating effort when they have greater mating opportunities (Marlowe 1999).

So male parental care is more complex to unravel than is maternal care, just as female sexual behaviour is more complex to unravel than male sexual behaviour, but there is nothing in forager societies that supports the equal sharing of sex or paternal care. From the female's point of view male parental care may appear 'fickle' just as female sexual responses may appear fickle to the male. How likely a male is to put his efforts into parenting compared to mating will be dependent on his situation (though we also would expect individual variation due to variation in inherited traits). If a male has multiple 'free' sexual opportunities we would expect his interest in parenting to be low.

We should also note that the 'situation-dependence' of female sexual behaviour includes the 'situation' of the male having resources in his possession from which the female can benefit. If females had those resources 'for free' this 'situation' and therefore this evolved female motivation for sex would be removed and potentially the number of sexual rejections of men would skyrocket. Ryan and Jethá argue, in contrast, that 'free' access to sex for males and 'free' access to resources for females would produce male motivation for parenting and female motivation for sex. The evidence strongly suggests they are wrong.

~~~~~~~~~~~

Ryan and Jethá (p. 99) use vampire bats as an example of sharing in nature: these bats will share blood with another bat that has been unsuccessful in feeding, and that recipient will return the favour on another occasion. True. That's reciprocity in food sharing and nothing to do with sex. These bats certainly do not share sex but live in groups of related females together with a few 'harem' males, and there are other groups of non-resident males. Males fight viciously for sexual access to the females (Wilkinson 1985).

(While on the subject of bats, studies have shown that bat species with promiscuous females, and therefore larger testes in males because of sperm competition, have relatively smaller brains than do species with females exhibiting mate fidelity. This pattern may be a consequence of the negative evolutionary relationship between investment in testes and investment in brains – both are metabolically expensive tissues so investment in one reduces investment in the other. See: Pitnick, Jones, and Wilkinson 2006).

Some sharing of food within extended family groups is easily enough accounted for by kin-selection and reciprocal altruism. Sex is another matter altogether. Apart from anything else, sexual reproduction is mostly about mating outside of the family; about mating with a non-relative, a 'stranger'. This non-kin aspect of sexual reproduction is both the attractant and the source of conflict.

Ryan and Jethá return again to the bonobo to argue that they have "sexual egalitarianism" and that we share some unique traits with them. As we have seen, bonobo sexuality is not actually as casual and "egalitarian" as people have been led to believe, nor is it useful to ignore the very many differences between bonobos and ourselves due to our very different selection pressures and evolutionary paths.

The authors (p. 101) argue that the abundance of sexual opportunity makes it less worthwhile for bonobo males to fight over any particular one female and that the motivation for male chimpanzees to form alliances to guard ovulating females "evaporates in the relaxing heat of bonobos' plentiful sexual opportunity". As we have seen, though, bonobo sex and "sex" are not in a relaxed atmosphere but one of tension over access to food, and it is often a way for incoming and other low ranking females to ameliorate the negative consequences of their low rank.

Extended receptivity – and we find receptivity rather than proceptivity (active solicitation of sex) in bonobos, while greater proceptivity is seen in female chimpanzees – does serve to reduce the intensity of male sexual interest because female attractivity is spread over a lengthier time rather than being focused on those rare and brief periods of actual ovulation. But this diffusion of female attractivity, receptivity, and male sexual interest can also serve a pair-bonding role if selection for male parental care is part of the package (more on this later).

Ryan and Jethá (p. 100) write that egalitarianism is found in nearly all simple hunter-gatherer societies because it offers the best chance of survival. They say that under the living conditions faced by modern hunter-gatherers and by all our pre-agricultural ancestors it may be the only way to live. They then explain this egalitarianism as the institutionalized sharing of resources and sexuality yet the latter is not what we find at all in hunter-gatherers today. The partible paternity societies they present as evidence are horticulturalists not hunter-gatherers but even there we have seen that their sexual behaviour is far from how it is presented to readers of *Sex at Dawn*. If we do look at hunter-gatherers we find monogamy and polygyny, some secretive extramarital sex and sexual jealousy, and the importance of marriage, sexual fidelity, paternity, and the sexual division of labour.

Ryan and Jethá (p. 101) ask, and it seems in all seriousness, "why presume the monogamous pair-based model of human evolution currently favored would have been adaptive for early humans, but not for bonobos in the jungles of central Africa?"

They are asking why pair bonds would have been adaptive for our human ancestors in the last million years or so when they were not adaptive for bonobos who were, as always, living in the forests. Our pre-human ancestors were most probably polygynous, as already discussed, and they continued to be so, with a mixture of polygyny and monogamy, for most of our history. Most human cultures still are. Even if they were promiscuous maters six million years ago there is every reason to believe selection acted against that during much of our evolution.

The question Ryan and Jethá ask could be replaced by multiple other strange questions of the same kind such as: Why presume that bipedalism would have been adaptive for early hominins (as it was) but not for bonobos?! Or: Why presume polygyny was adaptive for *Australopithecus afarensis* but not for bonobos?!

If we move forward in time to about one million years ago when the bonobos were starting along their separate evolutionary journey from the chimpanzee and our ancestors were now *Homo erectus* we could ask: Why presume the sexual division of labour would have been adaptive for our ancestors but not for bonobos?!

Bonobos did not evolve tool making or tool use, large brains, a sexual division of labour, fur loss, the need to carry offspring, human thermoregulation, males able to transfer between groups, female reproductive fat stores, and resource-hungry offspring with a very long dependency and developmental needs which could not be met by the mother alone. Are the authors seriously asking why two different species that share a common ancestor *six million years ago* can be presumed to have experienced different selection pressures and to have then evolved differently?

Ryan and Jethá (p. 101) argue that extended female sexual responsiveness in our ancestors would have enabled larger group sizes by reducing male-male conflict. Yet it has done nothing of the kind in bonobos where the males only ever associate with males in their natal group; only adolescent females can normally change residence, and in doing so they lose all contact with their natal group.

Extended female receptivity is not the means by which our ancestors became able to create extended networks; this required pair-bonding and in-laws. For this crucial aspect of human evolution we'll look at some anthropological evidence.

Evolutionary History of Hunter-Gatherer Marriage Practices

In humans but not in other apes, males can move away from their natal kin group. Bernard Chapais in *Primeval Kinship: How Pair-Bonding Gave Birth to Human Society* shows how stable breeding bonds, paternity recognition, and female transfer between groups paved the way for amicable relations between the males of those groups and therefore greater human social networks. Pair bonds meant that males not only gained connection to offspring but also to other adults in other groups – the relationships with 'in-laws' could only be recognized and sustained through the marital relationships of sisters and daughters (Chapais 2008).

Evidence supports a deep evolutionary history of limited polygyny and 'brideprice' or 'brideservice' that stems back to early modern humans and, in the case of arranged marriage, to at least the early migrations of modern humans out of Africa. Reciprocal mate exchange and then regulated marriages led to more complex between-group reciprocal alliances. Most cultures have rules concerning who their members can or cannot marry and these rules are connected to kinship-ties between groups. Real or classificatory cross-cousins are often favoured spouses and this emerges naturally as a result of multiple generations of sister or daughter exchange between two kin lineages. A key result of this is the affiliation of several unrelated males through each of their relations to the same female (e.g., related as a husband, father, brother, or brother of her husband). These relationships can ameliorate hostile relations across patrilineages and they also make possible the movement of people between different areas and therefore different resources (Walker, Hill, Flinn, and Ellsworth 2011).

It is now thought that flexible multilocal residence is the most common form of residence amongst foragers, with family units or individuals able to move to various different camps for varying amounts of time. These are often nuclear family units – two parents plus offspring – moving to live near relatives of either the husband or the wife. The forager diet requires mobility so having these kin ties on both sides increases residential options. Without long-term pair bonds this kind of bilateral kinship and flexibility of residence would not be possible (Marlowe 2004a).

Multilocality, bilateral descent (descent traced on both maternal and paternal sides), and paternal kinship ties are what a promiscuous mating system would not allow. Some sort of reduction of aggression between different communities of patrilocal males in our ancestry must have occurred, ultimately becoming flexible, multilocal living and bilateral descent. Pair bonds are a necessity for this to actually work. Humans have kinship ties through fathers as well as mothers, and it is 'marriage' which made this possible.

This argument has also been supported by others, such as Kim Hill who talked about his research in a *Science* magazine podcast[11]. For human males, those males who live in the neighbouring social group are often married to either a daughter or a sister or, in some cases, the mother. This means relations between the males can be friendly rather than hostile and creates a much larger social network. This also means ideas and information can flow from one group to another to another.

Hill also says that adjacent bands generally have peaceful relationships because they are exchanging marital partners. This is one of the big results from the research – all of the adjacent bands in a region are partially interrelated to each other because they are constantly exchanging marital partners and, as a result of those friendships, they are also visiting each other. When the hunter-gatherer bands do engage in warfare, which, Hill says, is quite frequent, it is with more distant groups that are not exchanging marriage partners with them. This warfare is often *particularly* to capture females as mates for the males in the group.

So here, rather than the very limited bonobo social cohesion, we have the extended human social networks where pair-bonding plays a

---

[11] *Science* magazine podcast interview with Kim Hill, *Science,* March 11, 2011, Vol. 331 no. 6022 p. 1340. http://www.sciencemag.org/content/331/6022/1340.2.full

crucial role. It goes a long way in answering the question put by Ryan and Jethá (p. 102) as to why monogamy is included as an explanation for human sociality: our males – like bonobo and chimpanzee males – would otherwise still be condemned for life to knowing no one and nothing beyond their small natal-group world if we did not have pair-bonding.

The authors (p. 104) again use a quote from E. O. Wilson (1978) who is indeed agreeing with them that understanding sex-as-bonding helps in our acceptance of homosexuality. But Wilson, as we have already noted, is not writing about multiple sexual relationships but only about *non-procreative* sex as being important to bonding. He is writing about this in connection to the heterosexual reproductive pair bond, facilitated by the human female's extended sexual receptivity, and including the man's natural desire for "exclusive sexual rights" over his wife. The authors consistently omit to say that E. O. Wilson is writing an attack on the Catholic Church's attitudes towards non-procreative sex within marriage and towards homosexuality. He is certainly not suggesting promiscuous sex-as-bonding is our 'nature'.

*

- There is a glaring disparity between Ryan and Jethá's presentations of sexual behaviour in the partible paternity tribes and what actually occurs.

- The authors have distorted information on what happened in Tahiti and failed to make clear the disparity between the fantasy and the reality regarding the Tahitian 'free love'.

- Women are commonly exchanging sex for resources, and accessing resources is a selection pressure for extended female sexual receptivity itself; if the resources came free it would remove a major evolved motivation for sex in females.

- By the same token, 'free' access to sex for males removes the selection pressure for male parental care of any offspring.

- Bonobo and chimpanzee males are indifferent to the resource needs of offspring.

- All human cultures have a reproductive sexual division of labour where males and females provide different things to each other and the family unit. Bonobos and chimpanzees do not.

- Sex is not shared in hunter-gatherer societies.

- The sexual promiscuity of bonobos acts against, not in favour of, both male parental investment and the formation of ties between philopatric communities.

- Our ancestors were not condemned to be forever in small isolated communities: pair-bonding broke down the barriers to extended human kin networks.

Ryan and Jethá close their CHAPTER SIX with their selected comments about the Mehinaku and the Canela which we have already looked at earlier in more detail. Other than picking out a few titillating titbits, Ryan and Jethá choose to give nothing more away about the lives of any of the South American tribes. They end their chapter saying (p. 103) that they don't want to risk overwhelming the reader "with dozens more examples of this community-building, conflict-reducing human sexuality".

Quite!

CHAPTER FIVE

Parenting and Marriage

In their CHAPTER SEVEN Ryan and Jethá continue with the theme of shared parenting, presenting more of their 'evidence' for our ancestral groups jointly producing children of unknown paternity and raising them communally with no bias towards genetic self-interest.

Human children need more than just the parental care from a mother to survive and prosper; no one would deny the importance and existence of this in our extended networks of family and friends. This care from people other than the mother – *allomothers*, who may be male or female – is a subject on which Sarah Blaffer Hrdy has focused (we have already met Hrdy in connection to her evidence regarding infanticide by males as a male reproductive strategy).

Ryan and Jethá (p. 106) quote from Hrdy's book *Mother Nature* (1999a) where she writes:

"Infant-sharing in other primates and in various tribal societies has never been accorded center stage in the anthropological literature.

Many people don't even realize it goes on. Yet...the consequences of cooperative care – in terms of survival and biological fitness of mother and infant – turn out to be all to the good."

We have already noted cases in monkeys, and possibly one in bonobos, where a high-ranking female will take the infant from a low-ranking female and due to neglect the infant will die (Silk 1980, Maestripieri 1993). Hrdy also writes that baboon and macaque mothers try to hold on to their own babies and that if it is taken by a dominant female the mother may not get her infant back and it will starve to death. So what is Hrdy referring to when she writes about infant-sharing?

The non-harming infant-sharing primates Hrdy does refer to are langurs and other leaf-eating monkeys which live in small one-male groups where the females are closely related. The allomothers in these species are young females who have not yet reproduced and who are curious about infants. The infants complain loudly when they are taken from their mother but getting some relief from the infant means the mother is able to feed herself better. The young females, though they can be rough with the infants, do not otherwise harm them. As the mothers are able to feed better their reproductive fitness improves – female reproductive success is very much about translating food into offspring, and variation between females in nutrition can lead to variation in reproductive success. The monkey females once they have produced their own infants quickly lose interest in the infants of other females.

Other primates where there is shared parenting are monogamous species such as siamang gibbons and titi monkeys where the male helps to carry offspring and the female can then feed herself more easily. Marmosets and tamarins produce twins and the male carries the infants except when the mother is nursing. Twins are an enormous reproductive burden on these females, and these families often have two breeding males with the one female plus other non-breeding helpers. This breeding system has been discovered to also include the fascinating mechanism of *chimerism* whereby cells move between the twins in the womb making their genetic make-up often a combination from more than one male. The effects of chimerism are still being studied but so far suggestions have been made for its selection in connection to the two-male breeding system and male parental care. Both sexes also appear to vary their care of offspring according to the genetic make-up of the offspring (Ross, French, and Orti 2007).[12]

Hrdy points out that in the other (non-human) apes and in most monkeys the mothers are totally devoted to their (usually) single offspring. One thing that separates human mothers – across all societies – from these other primate mothers is that if a human mother causes a death it is most likely the death of her own newborn. Human mothers far more than mothers of these other species make reproductive decisions based on the potential for any one infant to survive and prosper. This decision depends upon how much help may be available, what other resources are available, and how it will affect the prospects of the offspring she already has and those others she may have in the future.

Hrdy's (1999a, 2000) view is that fathers tend to be unreliable when it comes to providing for their children. If they can be relied upon then she argues that monogamy is the best option: a compromise between two parents where the children are the winners. The reason she says fathers cannot be relied upon is because of the possibility that they are not that infant's biological father. This means that men have a high threshold for responding to infants along with a low threshold for responding to reproductive possibilities with other females. A father's attention is therefore going to be far more focused on sexual opportunities than on the parenting needs of offspring.

Hrdy argues that for help with parenting a human mother most likely looked to her own kin: her own older children, her post-reproductive mother, and other similar female kin such as aunts. Hrdy does not make any argument or suggestion that men would reliably

---

[12] 'In the Marmoset Family, Things Really Do Appear to Be All Relative', Carl Zimmer, *New York Times,* March 27, 2007.
http://www.nytimes.com/2007/03/27/science/27marm.html

provide paternal care outside of monogamous relationships, and only sometimes, she argues, could they be relied upon within monogamous relationships. So Hrdy's 'shared parenting' is really an argument about men's lack of interest in parenting, and the mothers predominantly looking to close *female* kin for help.

The main problem with this argument is that it requires matrilocal residence for mothers to benefit from allomaternal care from female kin. We have already discussed the patrilocal residence of our ape cousins and how this is the most likely pattern in our own ancestors. If there was some change to matrilocal residence then that raises the question of why marriage became or remained so essential.

The current flexibility of residence is to take advantage of kin relations of the mother *and* the father, and the father's mother, sisters, and aunts can be parenting helpers when they know they are connected to the child through that male. We need to take into consideration the impact of these paternal female relatives, especially the father's mother, on children and, by extension, their impact on the sexual behaviour of the wife of a son or other close male relative. The influence bonobo mothers can have on their son's reproductive success reminds us that these patrilocal mothers evolve strategies for their own reproductive fitness through that of their sons.

As an evolutionary biologist Hrdy understands the conflict of interests between the sexes. She too, for example, writes about the toxic chemicals in the semen of the fruit flies (as mentioned earlier) and of the subsequent experiments where monogamy was imposed on these fruit flies: after 47 generations the semen of these males was no longer toxic, males treated the females better, and the reproductive output of the females actually improved. As Hrdy says, if such a thing could be imposed on other species, including humans, there is no reason to think that a similar result could not be obtained and that the priority of both men and women would be the well-being of children (Hrdy 1999a).

Hrdy writes that monogamy happens to be an unusually stable compromise, in part because it raises survival prospects of offspring. It "reduces inherent conflicts of interest between the sexes", and over evolutionary time, lifelong sexual monogamy "turns out to be the cure-all for all sorts of detrimental devices that one sex uses to exploit the other". But, as we know, sexual monogamy is rare. The alternative to monogamy, she says, is "for matters to sort themselves out so as to favour the sex with the most leverage, usually the males."

At first glance Ryan and Jethá's argument can seem like a solution: everybody just shares the sex and shares the parenting equally. But natural selection does not work like that. Perhaps it could be socially engineered? Russian revolutionaries in the town of Vladimir in 1917 had the idea of making every eighteen year old girl state property, registered at a bureau of free love. Men in possession of a certificate showing that they belonged to the proletariat would have the right to these women, and the children would become the property of the revolution. Though the idea was popular amongst the men it did not catch on because they could not agree on who should cover the costs of childcare (Jones 2002).

Is it really surprising that males are attracted by the prospect of 'free' sex but are far less keen on the subsequent *costs* of reproduction?

Our ancestral groups, like hunter-gatherers today, most likely had fluid membership comprising both close genetic kin and others who are related through marriage, and there was also the potential for reciprocal, mutually beneficial relations outside of family relatedness. But as we have seen, even in the South American tribes where people are particularly closely related there is still a bias towards those *most* closely related whether it is female kin sharing food and childcare within the immediate family or the men focusing their interests on children they believe are most likely their own. Beyond the tribe there is often great hostility and warfare. Genes for these traits cannot help but spread more easily than alternative genes that produce an indifference to genetic relatedness.

Ryan and Jethá (p. 108) quote from Janet Chernela (2002) who studied the South American Tukanoan, another tribe who believe in partible paternity. The quote refers to how the clan brothers provide for one another's children. So let's look in a bit more detail at what Chernela writes about the particular Tukanoan patriclan she studied: the Wanano.

The Wanano practice compulsory exogamy which means that the females do not marry within their own birth clan. The wives, usually cross-cousins of the husbands, marry-in from other non-Wanano clans and are very much viewed as 'other' and outsiders. This 'otherness' is made even more apparent because, due to the number of different languages and the marriage rules, the wives speak a different language in their birth clan than is spoken by their husband and his clan. Children are only allowed to speak the language of their father and his clan and they learn that they belong only to the father's clan, though daughters will leave at marriage.

The only potential non-incestuous sexual partners within the village are the in-marrying wives of the clan 'brothers'. These women do have discreet sexual liaisons with their husband's brothers and husband's brothers' sons, and the most common first sexual experience for a young man is with his father's brother's wife.

Women can deliberately humiliate and influence their husband by making a public display of their infidelity. One wife created a situation where she allowed herself to be discovered in a sexual liaison with her husband's nephew. A public argument between husband and wife ensued, and the husband also beat his wife. She accused him of not fishing enough for her and their children, and after the incident the husband fished for two days and nights. This is a strange accusation and a strange situation for a wife to set up if the fish the men bring in are really shared equally and regardless of paternity.

Another young wife had an affair with a man from a different patriline and in this case it led to a bitter and deadly feud because a potential biological father from outside the clan represents an intolerable breach of the descent line. Any tolerance of communal paternity only extends across the brothers of the patriclan; children are and have to be children of the patriclan – if the mother leaves the clan the children have to stay behind.

Few marriageable partners are actually available to any man, and those that are available are the same few that are available to his brothers. This means, Chernela says, that there is an ambivalent nature to the relationships between brothers who have to show solidarity while at the same time they are in competition for the same female cousins.

So, even in groups like these with close bonds – including genetic ties – between the males in the all-important patriclan there is still little openness in extra-marital sexual behaviour within the clan. These are peoples who, if Ryan and Jethá are correct, should have no reason for marriage at all. And why should a wife publicly humiliate her husband and suffer a beating to manipulate him into fishing more when all are supposedly provided for equally?

The total unacceptability of women having sex outside of the patriclan also shows that rather than Ryan and Jethá's vision of sex leading to the improved relations between different groups we find instead the complete intolerance of such sexual contacts. This is just another example where the 'evidence' for an attractive and sexy picture of shared paternity loses its shine when more details are revealed.

Nuclear Meltdown?

Ryan and Jethá (p. 110) argue that "untold numbers of infants whose existence threatened to expose the colossal error at the heart of [the nuclear family] were being sacrificed, quite literally, in foundling hospitals" and (p. 111) "children whose existence might have raised inconvenient questions about the 'naturalness' of the nuclear family were disposed of in a form of industrialized infanticide." Strong words, so let's look a bit more at these foundling hospitals.

Foundling hospitals were set up across Europe from the 15th century because of the numbers of babies abandoned along roads and in gutters, and who were therefore being condemned to a certain death. The foundling hospitals were initially an attempt to save some of those lives but their existence then meant that hard-pressed parents became more likely to abandon infants to these institutions where they might have a chance of survival. These were often children of married couples too, not just of unwed mothers. Parents who would not have left their infants to a certain death in the gutter now felt able to leave them at the foundling hospitals where the parents could at least imagine they would be cared for. But this great increase in numbers of abandoned babies meant that soon there were too few wet-nurses to feed them so death due to hunger and disease ensued (Hrdy 1999a).

This was nothing to do with hiding any failings of the nuclear family. Ryan and Jethá's suggestion here is that with communal sex and communal paternal investment these children would have survived and prospered but, as we'll see in Chapter Seven, their later argument is that communal parenting and the hunter-gatherer 'good life' actually requires the disposal of any infant that would increase the population. That means no more than an average of two children per woman would be allowed to live in order to keep the population stable and in tune with resources. Their own communal parenting argument, therefore, is for the disposal of any infant beyond the population replacement level.

Ryan and Jethá end their CHAPTER SEVEN (p. 112) with a quote from Laura Betzig where she calls Lewis Henry Morgan's group marriage ideas fantasy. They say that Morgan's understanding of family structure was no "fantasy" but based upon decades of extensive field research and study. Yet that is their last mention of Morgan, and apart from stating Morgan's view that primitive promiscuous sexuality had to have existed, readers are not given any information about Morgan's research findings or any explanation as to why his group marriage ideas have been, if Ryan and Jethá are correct, *mistakenly* dropped.

It was not because of some morality issue that Morgan's ideas fell out of favour; he was, after all, arguing about the evolutionary improvements from a 'barbarous' primitive promiscuity to the 'superior' condition of marriage and the nuclear family. It just turned out that his interpretation of what kin terms tell us about our past was mistaken and his ideas on ancestral group marriage and other levels of human 'progress' have failed to stand up to scrutiny.

As we have seen, our ape cousins have patrilocal residence which almost certainly was at our root too; any matrilocal residence in our evolution would have come about later and it has never been the most common form of residence. We have also noted that Morgan assumed incestuous sex at our very origins, at least between brothers and sisters if not also fathers and daughters, whereas we know this does not happen in other species and that in our ape cousins females disperse.

We have already noted Morgan's ideas coming from kinship terms where, for example, all same-sex members of a generation may be classified under the same term which was mistakenly taken to be (at least in their origins) equivalent to our descriptive use of, for example, "father" and "mother". Though a number of societies have these shared kinship terms there is a clear understanding by the people involved who the mother and father actually are or, sometimes in the case of the fathers, most likely to be. Within a small group everyone knows which woman gave birth to which child and with whom she was having sex. The classificatory terms that are applied to groups of people of particular generations, and other subsets of relatedness, serve to tie the whole group together in relation to each other and differentiate that group from other groups.

According to Ryan and Jethá our ancestors simply all had sex with each other and that is all we need to know to explain why sexual monogamy is so difficult for us today. Or they may be saying (it is far from clear) that our ancestors had sex with a number of, rather than all, others. This immediately raises the question of 'mate preferences' and why they should exist at all. If they do exist then we immediately have the problems of who prefers whom, who is chosen, who is rejected, and who gets jealous. The authors have failed to consider the problems that arise from mate preferences, apart, perhaps, from their suggestion that sex would have been enforced when at least one of the participants (most likely the female) is being too 'stingy with the genitals' for the good of the group.

Ryan and Jethá avoid dealing with questions arising from mate preferences and their argument implies that everyone finds everyone else equally attractive, which, of course, never happens in the real world and is an intrinsic part of evolution by natural selection. They primarily present us with a limited picture of a promiscuous bonobo-like ancestor and tell the reader that this bonobo-like ancestor persisted until agriculture descended on us to spoil the party.

The evidence says otherwise.

One rather significant piece of evidence is that every hunter-gatherer group ever encountered already had marriage, i.e., they already had identified reproductive and economic bonds between a male and one or more females. Sex, of course, does not only occur within marriage but marriage is a very strange thing to exist in every remote foraging group (as well as everywhere else) if it only came about a few thousand years ago in areas where people started to settle. Ryan and Jethá offer us no explanation for this glaring elephant in the room.

~~~~~~~~~~~~~~~

In their CHAPTER EIGHT Ryan and Jethá continue to question the universality of marriage and they start by throwing in (p. 115) a mention of Robert Trivers' 1972 paper *Parental Investment and Sexual Selection*, saying that this paper "consolidated the position of marriage as foundational to most theories of human sexual evolution".

We have already noted Robert Trivers' paper, which has no mention of humans, in Chapter One. To recap, rather than it being males more eager to mate than females because males make many tiny sperm but females invest so much more in each egg, Trivers showed that it was parental investment that was the crucial factor. It is through this insight that we can understand sex-role reversed species where the males invest more in parenting than do the females and it is therefore the males who become the limited resource for which females compete. The sex with the largest parental investment is a limiting resource for which members of the other sex compete (Trivers1972).

In their note to their mention of Trivers, Ryan and Jethá say that his paper is seen as a foundational text in establishing the importance of male provisioning as a crucial factor in female sexual selection. Well, yes: females can out-reproduce other females if they obtain more resources from males that can be translated into more offspring, so male provisioning, where it arises, will become an object of female mate choice. And not simply female mate choice: where male parental investment is important for offspring survival then females will also be competing with each other for those resources and males will be making mate choices too. Without expansion from Ryan and Jethá on their objections to this evolutionary mechanism it is impossible to know what the authors are concerned about here other than this going against their agenda of 'free sex' for men and their promotion of females as natural 'sluts' rather than natural 'whores'.

Primatologist and anthropologist Meredith Small is quoted a few times by the authors in respect of her studies of primate females and their less than sexual monogamy – something that is not in dispute. In *Female Choices* Small (1993) notes the important difference between 'choice' in terms of a preference for a particular mate and 'choice' in terms of who actually is 'chosen' (i.e., mated with) in reality when the self-interests of others get in the way of those sexual preferences. And after her evidence for polyandrous mating in primates Small suggests that human marriage probably arose to accommodate dependent human offspring, and she concludes that our reproductive compromise has been the evolution of a human pair bond that isn't necessarily exclusive.

Female primatologists, anthropologists, and evolutionary biologists have rightly been keen to show how human females are not 'naturally' the sexually monogamous and passive reproductive vessels for use by males. Females too have reproductive strategies and seek to make their own active choices in their lives. The reason these writers do not suggest the existence of a casual sharing of sex, group marriage, and shared male parental care is because they understand more of the evolutionary biology involved and they recognize this 'free sex' is not going to be selected for in females.

Males of most species are usually more focused on 'mating effort' rather than 'parenting effort'. If males can get others to carry the costs of their offspring they may well choose to use their own resources to increase the number of sexual partners they can get. As Sarah Hrdy writes (2000):

"[W]hat impresses behavioral ecologists like Kristen Hawkes, James O'Connell, and Nicholas Blurton-Jones is how often men today (in the case of their studies of hunters among African foragers) seem more interested in maximizing prestige (which translates into more sexual partners) than in actually provisioning families. They emphasize the impracticality of foraging decisions, particularly men's obsession with large and prestigious prey like eland, even when higher returns in terms of protein can be obtained by targeting more abundant, but less prestige-enhancing small prey."

The survival of human offspring has no doubt required different strategies in different environments or combinations of circumstances but it did not evolve to encompass equally shared sex and equally shared male parental investment. Our extended social networks across more and more groups of people did not come about by sharing sex and paternal care across these increasing numbers of people but by the connections created by pair bonds between one man and one woman (more connections for the men in polygynous marriages). These connections with both blood and marital kin made the mobility of reproductive and economic units possible throughout the kin network that marriage created.

~~~~~~~~~~~~~~

It is true that 'marriage' can take various forms in different human societies but the fact that it exists in some form everywhere leaves us with the question of why something that is argued by Ryan and Jethá to be so unimportant is, in fact, the opposite. Why do all of the societies we have looked at concern themselves so much with marital bonds? Sex, of course, is not synonymous with marriage, but in the group sex scenario of Ryan and Jethá marriage ought to be redundant rather than merely variable.

The Aché, Ryan and Jethá (p. 119) are keen to tell us, can change a spouse with relative ease; they do not add, though, that to be without a husband means a mother might as well kill her own children (and others are not averse to doing the killing) because their prospects are so poor. As we noted from Hrdy (1999a) earlier: "Even if the father was still alive, Aché children whose parents divorced were *three times more likely to be killed* than if the marriage endured" (emphasis in original).

For the Curripaco, Ryan and Jethá say that marriage is a gradual, undefined, process. They quote from Valentine (2002) that younger Curripaco consider a couple who merely hang their hammocks together as married, though older Curripaco disagree, saying that only when the couple have demonstrated that they support and sustain each other are they married. Traditionally the birth of a child and the two parents showing their commitment through the joint fast cements the marriage, i.e., a successful reproductive unit with an interdependency of the two parents has been established. The Curripaco, as noted earlier, also have strict rules against sex before or outside marriage, and people cannot subsist without a spouse due to their economic interdependency.

Babies born outside marriage are likely to be starved to death. Marriage, therefore, is hardly undefined as Ryan and Jethá argue.

It is hard to understand how Ryan and Jethá continue to 'miss' this kind of information in their own sources. Marriage may be viewed differently and applied differently in these societies but it is crucial all the same and includes common features. The authors of the *Sex at Dawn* story only seek to imply relaxed and open attitudes towards sex and marriage which in reality do not exist.

Ryan and Jethá mention the !Kung, attempting to astound the reader with the information that "most girls *marry* several times before they settle into a long-term relationship" (emphasis in original). So let's look at this in a little more detail.

In previous generations first marriages would be of girls as young as ten or twelve though without sexual relations until the girl reaches menarche. Husbands are 20-30 years old and first have to kill a large animal to be eligible. When the girl is still young these marriages are arranged by others and they are often unstable until she starts having children with one, usually life-long, partner. The young girls may be pressured into accepting these early marriages as they are important to older family members in creating ties between in-laws and for the 'bride-service' the husband provides (Shostak 1990).

Young !Kung girls are usually afraid of their new husband, and an older woman will sleep between the couple until the girl gets used to being with the man. The early years of marriage can be stressful and divorce is not uncommon though emotions will still run high.

All we have here is another version of marriage but with many of the same components including parental responsibilities, fathers as providers, the importance of ties through in-laws created by the marriage, and strong emotions connected to these relationships.

Are we still astounded that they marry several times? Is this pointing to a bonobo-like, shared sex ancestry for our species?

The Canela are mentioned again (p. 120) with regard to virginity. Ryan and Jethá say that virginity is so unimportant that there isn't even a word for it in many societies then they say, as if offering an example, that virginity loss is only the first step to marriage in the Canela. In fact, for the Canela, virginity is *so important* that, for both sexes, masturbation is strictly forbidden. A girl is warned that she might lose her virginity payment if the hymen is stretched or broken. Boys are warned that handling themselves may loosen their foreskin and thereby cause the loss of their virginity payment which would become generally known and would be an embarrassment to their family (Crocker and Crocker 1994).

Ryan and Jethá then go on to mention the several steps the Canela girls go through before they are properly married. I have already written at some length about these steps the girls go through, including the sequential sex occasions (discussed in Chapter Four). The authors mistakenly write that the girls earn payments of meat paid directly to the girl's mother-in-law on a festival day, confusing the festival sequential sex with a different occasion where the girl goes with one male on a hunt and may or may not have sex with him but she takes the meat he provides to her mother-in-law as part of her greater acceptance as a daughter-in-law.

We have already noted that the Canela girl wins her maturity belt only after the sequential sex service which, after a brief period of seclusion, is painted by her female in-laws. As we have seen, by the end of their teenage years girls will likely have become mothers and their life changes dramatically to one of childcare and domestic work. It is not unusual in many societies for marriage to be only fully recognized with the birth of a child, after all, that is its main purpose for all those involved in the union which joins two families together through a shared interest in the same children. It is not sex that creates links between unrelated or distantly related – or even fairly closely related – groups, but reproduction.

Of the Warao, Ryan and Jethá (p. 121) write of *mamuse* relations and that: "During these festivities, adults are free to have sex with whomever they like."

No.

In this fertility ritual men might agree to exchange wives and, upon payment of a substantial price called *horo amoara*, "skin payment", they are free to engage in dancing and sex (Olsen 1996).

This is the men using sexual access, for a price, to their wife which can lead to a mutually beneficial lifelong friendship between the two men and has a potential 'insurance' benefit for their offspring. Outside of this use for male-male alliance, Warao men do not like their women to have affairs, and scenes of jealous quarrels are quite frequent. Sometimes a child is referred to disapprovingly as *dimamana*, "of two fathers". Only if a man is seriously ill might he voluntarily give his wife to a brother or cousin (Heimen 1997).

Ryan and Jethá comment (p. 121) that the Pirahã people keep their gene pool fresh by "permitting their women to sleep with outsiders". This actually refers to the fact that although this small group of people have kept themselves mostly unaffected by outsiders, of whom they have a low opinion, they do have tools and consumables from the outside world because it is not uncommon for Pirahã women to trade sex for various items from the boats that frequently come by. So this is the usual use of sex by women in exchange for other resources that is enabling some gene flow into the Pirahã from the outside world.[13]

Then Ryan and Jethá tell us that it is common among the Siriono for brothers to marry sisters, "forming an altogether different sort of *Brady Bunch*". They say: "Just rehang your hammocks next to the women's and you're married, boys", which makes it sound as if a group of brothers move their hammocks next to a group of sisters and they are then all married as a group.

No.

The Siriono band numbers 60 to 80 individuals and they live together in the same house. Marriages between a man and his mother's brother's daughter are preferred which means that in a small inbreeding group like this brothers will have the same potential wives

---

[13] From a talk with Daniel L. Everett for *Edge*, retrieved August 7, 2011
http://www.edge.org/3rd_culture/everett07/everett07_index.html

who will be sisters and vice versa. Marriage merely entails telling the parents-in-law of the decision and the man moves his hammock from near to his parents to near to his wife's parents. Monogamous, and sometimes polygynous, nuclear family groupings in the communal house make up the band. The nuclear family is the fundamental social and economic unit among the Siriono (Holmberg 1969). We will meet the Siriono again in the next chapter when Ryan and Jethá refer to them in the context of jealousy.

As for spouse exchange in the Inuit, which the authors add as another example of a "casual" approach to marriage, this has indeed been an important part of their culture. It is not to be entered into amongst relatives or *within* any group but is used between couples from more distant communities due to the potential for a stranger coming upon another community to be killed. By creating these links this killing could be avoided. Wives, though, could not be exchanged between known enemies. Warfare was common and women were sometimes captured; if raped they were killed soon afterwards to prevent the creation of half-siblings. One man who discovered his wife had been having sexual relations with his trading partner without his permission "stomped" his wife to death then started to go crazy and was executed (Hennigh 1970).

There is certainly no casual attitude towards sex amongst the Inuit. In two known cases of an Inuit woman becoming pregnant before marriage, one starved herself to death before the birth and the other kept the pregnancy secret then buried the baby. The threat of jealousy is certainly recognized and spouse exchange is not entered into lightly but it is a mechanism that works to avoid inter-group killings in a people who must travel widely in a region where people are widely dispersed. It results from the need to make links with those people who in other societies would simply be strangers, and therefore potential enemies, that could be avoided. And like so many of these instances where men use their wives to create beneficial links for themselves to other men, the wife's perspective seems irrelevant.

Ryan and Jethá then (pp. 121-122) mock a couple of 'moral guides' from history that advised against too much sex, even with a spouse, but restrictions on sexual activity is something which is not uncommon amongst many tribes, including those with partible paternity. Many of these tribes believe that sex reduces a man's strength so abstention is a way to build strength. The Canela have sex restrictions in male adolescence for up to three years depending on how determined the boy is to become an enduring runner and skilled hunter. Numerous tribal societies have taboos against sex after the birth of a child or when a child is ill or on other occasions when sex is viewed as polluting or otherwise potentially dangerous. The idea that cultures that have different ways of marrying somehow have a more casual attitude to sex turns out time and time again to be wrong.

Paternity Certainty Still

Ryan and Jethá begin their CHAPTER NINE with a quote from Robert Edgerton (1992) about the Marind-anim and their former treatment of brides on their wedding night. This involved as many as ten men from the husband's lineage having sex with the bride, with more the following night if there were more than ten men in the husband's lineage. Ryan and Jethá add no more to what they merely refer to as a "wedding celebration" and "unchaste shenanigans".

What Edgerton also writes is that semen was imagined to be very powerful by these people who also practiced ritual male homosexuality in the belief that it would aid a boy's growth. It was believed that filling the bride with semen would assure her fertility; the young bride, though, would be left so sore that she could hardly walk.

The Marind-anim were ferocious head-hunters. They would even raid enemies as far as a hundred miles away and capture children to raise as Marind-anim – their own numbers were dwindling; their women, it turns out, were largely infertile. The irony is, Edgerton writes, that their infertility appears to have been *due to* the filling of those women with semen from so many men which was leading to severe pelvic inflammatory disease.

Ryan and Jethá also write (p. 125) about the Kulina where the men are motivated to hunt by a woman other than a wife offering herself as a reward. The Kulina we have already expanded on in the previous chapter. Their population is very small – traditionally villages would only number 20 to 50 in total, all sharing the same longhouse – and so closely related that they are simply divided into those who can be sex partners and spouses, and those who cannot. Because most of the meat goes to the men many children have chronic protein malnutrition once past weaning and child mortality rates are about 40%; it seems that even this use of sex by the women as an inducement to hunt does not get enough protein into the mouths of their children. And even though there is this acceptance of women exchanging sex in the hope of getting more meat there is still much discomfort amongst the men at the thought that a brother may be the father of his child (Pollock 2002).

These examples do indeed show how differently societies manage their sex and reproduction but it is only ever a variation on the universal theme of marriage, women exchanging sex for meat or other resources, a sexual division of labour with interdependence of spouses, low offspring survival, kinship ties between some people often along with hatred of those not related or more distantly related, variation depending on patrilocal or matrilocal residence, conflict between the reproductive interests of the sexes, and so on.

The Mosuo

Ryan and Jethá think that the Mosuo is a society which undermines the evolutionary theories that place sexual pair-bonding and paternity certainty at the centre of our evolved nature. They present a rosy picture of casual sex between men and women, mostly using quotes from Yang Erche Namu's *Leaving Mother Lake* and Cai Hua's *A Society Without Fathers or Husbands*.

It certainly can seem like a good solution to the problems of rearing human offspring and male-female parental conflicts: take the father out of the equation, give the mother's brothers the male parental role, and keep maternal kin together to provide the family and economic base. For most outsiders there is an image of secretive, nightly sexual encounters between men and women and a gentle and happy daily life with kin. Before coming to any conclusions about what this says about human nature we need to have a few more insights into the workings of the Mosuo society. Using the same sources as Ryan and Jethá, and adding to them from a few more, we'll now look at the Mosuo in some detail.

Traditionally about 10% of Mosuo were cohabiting couples and another 10% engaged in a form of marriage. Fathers are usually known and it is not unusual for a father to help the family of the mother of his child during busy agricultural seasons (Shih 2010).

Ryan and Jethá (p. 129) write that particularly libidinous women and men unashamedly report having had hundreds of relationships, but they provide no reference for this. Anthropologist Tami Blumenfield writes that some people may have had 30, 40, or even 50 partners over a lifetime. Blumenfield has written a fact sheet to debunk some of the myths about these people, including the one that they do not know their fathers: women, she reports, are embarrassed if they cannot name the father of a child.[14]

What Ryan and Jethá's own source Cai Hua writes is that there is in reality a limited number of potential sexual partners which makes the authors' "hundreds of relationships" very unlikely. These people live in small scattered villages amongst blood kin. Young men may go from village to village trying to be accepted by a woman but as this

---

[14] *The Na of Southwest China: Debunking the Myths,* Tami Blumenfield, 2009 http://web.pdx.edu/~tblu2/Na/myths.pdf

searching for sex takes place at night, travelling far and then returning for work the next morning is not an easy option and is not something older men are able to do (Hua 2001).

To illustrate, Hua describes one typical village and the potential number of these *açia* relationships a man or a woman *might* have which is between five and twelve. The actual number is even less when other factors involved in making a choice were included. For the 13 women and 12 men Hua used as an example, there was an average of five açia relationships for women and six and a half relationships for men.

Sexual activity is accepted between the ages of thirteen (though rarely so young) and fifty for females and between the ages of thirteen (though none known at that age) and sixty-one for males. Sex beyond these years – 50 for women and 61 for men – is frowned upon. The most frequent visiting is up to the age of thirty and then visits become less frequent. People then tend to see the same sexual partner more often than they seek out new ones.

Hua describes how there might be two or more men at a woman's door trying to convince her to open it when other men will arrive and join in the attempt. In the end one man will persuade the woman to let him in. Men who travel to another village just because of a sexual interest in a woman will be chased off with insults and violence from the men there. And women must not take the initiative and visit men or they risk being labelled "a sow in heat charging through the fog".

Couples who make their relationship public spend time together in the daytime and call each other 'friend' or 'partner'. If someone has furtive sex with someone in one of these 'open' (i.e., publicly acknowledged) relationships it is called 'stealing sex', and a man caught stealing sex in this way is likely to be beaten by the woman's partner. The few men who manage to have these publicly acknowledged relationships with women in different villages are normally men who are rich and generous to the women.

The break-up of an 'open' relationship is usually by the man when he becomes attracted to another woman. Some women find the rejection difficult to accept and such a woman will try hard to convince the man to take her back.

Some conflicts due to sexual jealousy have resulted in serious injury and even death.

In *Leaving Mother Lake* (Namu and Mathieu 2004) we learn that some of the 30,000 Mosuo are actually patrilineal with contract marriages. In those that do have the 'walking' sexual relationships people are not entirely immune to jealousy and heartbreak but these feelings are greatly discouraged.

Christine Mathieu wrote this book with Namu about her life and she has studied the Mosuo more than most. Mathieu writes that this is not a matriarchal society: women are in charge of the household and men of outside affairs. She says that from an outside perspective:

"[I]t is difficult not to notice that male occupations are highly valued, that women shoulder a greater burden of physical labour than men, and that men command respect and authority more, because of the aura of knowledge they carry with them from their activities in the outside world."

Also:

"What is beyond argument...is that Moso society is not ruled by women as is invariably publicized by the mass media. Before the Communist revolution the Moso were governed by male chiefs who inherited their position from their fathers and passed it to their sons... Today...the administration is dominated almost exclusively by male cadres. Unlike women, who are constantly preoccupied with housework and farmwork, men are available to pursue positions in the outside world, to become village chiefs, administrators, cadres, technicians, teachers, traders, they have a fair share of authority in public and family life. Of course, in the Moso family, the maternal bond determines blood ties but this makes Moso society matrilineal, not matriarchal."

Mathieu also writes that there is no doubt that the Mosuo feudal rulers encouraged both the matrilineal family system and the custom of visiting relationships. This is a point that is also made by the *Lugu*

Lake Mosuo Cultural Development Association who write that historically the Mosuo actually had a feudal system in which a small nobility controlled a larger peasant population. The Mosuo nobility practiced a more traditional patriarchal system which encouraged marriage (usually within the nobility), and in which men were the head of the household. They state:

"It has been theorized that the "matriarchal" system of the lower classes may have been enforced (or at least encouraged) by the higher classes as a way of preventing threats to their own power. Since leadership was hereditary, and determined through the male family line, it virtually eliminated potential threats to leadership by having the peasant class trace their lineage through the female line. Therefore, attempts to depict the Mosuo culture as some sort of idealized "matriarchal" culture in which women have all the rights, and where everyone has much more freedom, are often based on lack of knowledge of this history; the truth is that for much of their history, the Mosuo "peasant" class were subjugated and sometimes treated as little better than slaves."[15]

This association also states that while promiscuity is certainly not frowned upon like it is in most other cultures, most Mosuo women tend to form more long-term pairings and do not change partners frequently; it is a system better described as one of "serial monogamy". Women can change partners but tend to do so relatively rarely, and while with one partner they will rarely invite another. Many Mosuo women have had a "walking marriage" relationship with the same man for twenty years or more. Also, if a father does want to be involved with the upbringing of his children he will bring gifts to the mother's family and state his intention to do so.

What's more, while Ryan and Jethá (p. 127) say that the Mosuo have no words for *murder* and *rape*, the *Lugu Lake Mosuo Cultural Development Association* state that, while this is technically true, murder and rape certainly do happen. They state:

---

[15] The Lugu Lake Mosuo Cultural Development Association http://www.mosuoproject.org/matri.htm

"There is a great danger for people coming in to study the Mosuo to idealize their culture; and the Mosuo themselves will tend to encourage this, as they don't like to talk about such things with outsiders. It is important to remember that, while the Mosuo culture is certainly fascinating, and has many aspects from which other cultures should learn, it is a *disservice* to them to describe them in a manner which is untrue.

"As with every culture, the Mosuo culture has both its good and its bad points. Those who are truly interested in the Mosuo should reflect both sides honestly, rather than distorting the Mosuo culture to support a personal agenda."

Quite.

What is clear is that the Mosuo are not some isolated remnant of a promiscuous pre-agricultural way of life. As well as the influence of politics and class conflict in the encouragement of matrilineality, another factor which would have encouraged this system is that the men traditionally engaged in long-distance trade and could be away from home for months at a time. This clearly strongly favours leaving the running of the farm work as well as the housework to women, and makes male control of female sexual behaviour difficult. The matrilocality and matrilineality here is a solution to particular circumstances, circumstances that were not experienced by hunter-gatherers. It is a modern system.

In many ways, though, it is a system that does have its advantages. Brothers and sisters, rather than husbands and wives, can provide the two economic roles of the two sexes. The men know they are biologically related to their sisters' children so providing for them is not hindered by potentially not being related. The brothers and sisters are also, of course, closely biologically related which has advantages over the more distant, if any, relationship between husbands and wives, though it also means that sex talk and flirting is absent from daily life – quite inappropriate behaviour when the only members of the other sex present are siblings. Other men are available to provide the sperm for reproduction, so the question is: why do we not see this system far more often?

As already noted, this is not how our ancestors would have lived in their mobile, hunter-gatherer groups. The settled matrilocal societies most likely arose in people where the women did the farming or horticulture and passed this land to daughters. In the Mosuo many of the men would also have been travelling with their trading caravans for long periods of time. That the Mosuo stand so alone in their replacement of husbands by brothers does suggest that there are strong forces working against this extensive severance of links between mothers and fathers.

Hua asks why other matrilocal societies normally still have marriage when, as we see in the Mosuo, it is not necessary for reproduction or the economic division of labour. He sees the answer in the human desire to both possess one's partner *and* to have multiple partners. Even in the Mosuo these conflicting aspects of human sexuality are still there. Hua concludes that the absence of marriage as the norm in the Mosuo make them the exception that proves the rule.

One further point of particular interest is that Namu says many Mosuo women cannot conceive. The fertility rate has been consistently lower than neighbouring groups over their known history. A study by Shih and Jenike (2002) looked at this depressed fertility in the Mosuo and found a relatively late age at first birth (median 23 years), long interbirth intervals (median 3 years), and a high rate of childlessness (16%). They concluded that this pattern of low fertility is an outcome of the unique Mosuo cultural practices. In addition, for women born between 1905 and 1929 pathological sterility caused by sexually transmitted diseases is likely to have also depressed fertility.

Though sexual activity begins in a girl's late teens or early twenties, these tend to be brief sexual relationships, or may not include sexual intercourse because the girl's sexual autonomy means she can move at her own pace. Because she is not in a marriage she is under no pressure from a husband to have sex so the actual frequency of sexual intercourse may be so low that the likelihood of conception is reduced.

The household also only requires a limited number of sons and daughters, and to produce too many offspring is a threat to its harmony. The absence of marriage also allows the women to take their time before resuming sex after a pregnancy and so they can space their births wider than if they were subject to a husband's sexual interest.

Pelvic inflammatory disease, most often secondary to repeated infections with sexually transmitted diseases, is thought to have affected women born between 1905 and 1929 due to the opening of trade routes, including the opium trade. Syphilis became a serious problem in the 1940s and 50s and was treated with penicillin. This was a period of particularly depressed fertility with the inflammatory response to STDs causing sterility in both sexes.

To go back to primates in general for a moment, there has been a study of promiscuity and the primate immune system which suggests that the risk of sexually transmitted diseases has been a major factor in the differences in primate immune systems: a higher white blood cell count was found in species where females mated with more males. Humans were found to have counts more consistent with monogamy than promiscuity, aligning most closely with the polygynous gorilla, and secondarily with the monogamous gibbon (Nunn, Gittleman, and Antonovics 2000).

Over the long term Shih and Jenike conclude that low population growth among the Mosuo is primarily a result of cultural practices. This is due to the low demand for offspring amongst co-resident sisters and the autonomy the women have over their own sexual activity. This is an interesting aspect of female sexual autonomy. It is assumed by Ryan and Jethá that the removal of social constraints on female sexual behaviour will lead to an increase in sexual activity; this evidence from the Mosuo suggests the opposite.

What needs to be kept in mind is that female mate choice is at least as much about being able to say *no* to sex as it is to say *yes*. In marriage contracts there is the implicit or explicit 'right' of the husband to the sexual and reproductive 'use' of the wife. The removal of marriage in the Mosuo removes this 'right' and appears to have led to a decrease in sexual activity.

~~~~~~~~~~~~~~~

The Mosuo are not a matriarchy. Ryan and Jethá put this argument down to terminology and they say that the real problem is in trying to limit 'matriarchy' to a mirror image of 'patriarchy'. But the absence of patriarchy does not a matriarchy make. If 'matriarchy' does not mean a mirror image of 'patriarchy' then another term is surely required.

Ryan and Jethá (p. 133) quote from Peggy Reeves Sanday who, on asking which sex rules among the matrilineal Minangkabau, was told:

"Neither sex rules...because males and females complement one another."[16]

If neither sex rules then it is neither patriarchy nor matriarchy. In the matrilineal Minangkabau, as in many matrilineal societies, household authority ultimately is in the hands of the eldest brother and passes via the sister to her son. They are also settled, land-working people so not representative of our pre-agricultural ancestors.

The problem with 'complementary' sex roles is that this applies to all traditional societies because they all have a 'complementary' sexual division of labour even if what each sex does varies across societies. Women do more of the work centred more on the home and children while men do more of the public activities and public decision-making; this is something seen in all societies from the Mosuo, through all the tribes we have looked at, to the Western 'ideals' of family life of the 1950s.

Some of the most sexually egalitarian societies, such as the African Aka people, have more shared activities of the two sexes (men

---

[16] http://www.eurekalert.org/pub_releases/2002-05/uop-imm050902.php

and women hunt together, for example) within the context of a strongly bonded nuclear family. The men of the Aka are described as the best fathers in the world, and there are very strong bonds between husband and wife, and between father and child (Hewlett 1991).

Ryan and Jethá (pp. 132-133) argue against Steven Goldberg's attack[17] on Sanday's use of the term of 'matriarchy' to describe the Minangkabau. Goldberg argues in general terms that the problem for women is not 'patriarchy' but the lower status given to traditional female roles – roles that are clearly valued in the Mosuo and the Minangkabau. The central role everywhere, Goldberg says, will forever belong to women because of sex, reproduction, and motherhood. He too argues that the sex roles are complementary and, more than that, he argues that men are more dependent on women than women are on men (Goldberg 1993).

If Sanday and Goldberg are, as it appears, both arguing that complementary sex roles are a levelling mechanism then we have to ask what the disagreement is about. We might even ask, with 'complementary' sex roles across societies, what actually *is* the problem? Goldberg's answer is that the real problem is the lower status given to female roles, so is the answer about elevating the status of female roles, as both Goldberg and Sanday argue, or is it about females taking on more traditional male roles from which they have been excluded, and men taking on more of the traditional female roles? Are different but complementary sex roles synonymous with sexual equality or not?

Peggy Reeves Sanday appears to be in favour of elevating traditional female roles to the *dominant* position and she sees the Minangkabau maternal role as the role that is viewed positively as most virtuous and closer to nature. This is why she argues it is a matriarchy – *maternal* values dominate the whole of society, she argues.

---

[17] In Goldberg's book: *Why Men Rule* (originally *The Inevitability of Patriarchy*)

Sanday quotes a Minangkabau belief: "Women *must* be given rights *because* they are weak. Young men *must* be sent away from the village to prove their manhood so that there will be no competition between them and their sisters" (emphasis in original).[18]

There are important leadership roles for fathers and uncles in the home but men – and traditionally not women – pursue more political and intellectual roles, often spending many years travelling away from their homes and own people. The 'matriarchy' that Sanday is arguing exists is one where the sexual division of labour is much as we see everywhere *but* the domestic sphere of women and children dominates and is at the centre of their value system. This is not exactly what Western feminists have been fighting for when seeking a greater presence for women in the public sphere and a greater role for men in childcare and domestic responsibilities.

Like Ryan and Jethá I too find much to argue with in Goldberg's views but there are still important points here regarding the differences between the sexes and the relative status given to those traits most associated with females and those most associated with males. The only real difference between Goldberg and Sanday is that Sanday argues in support of the *dominance* of maternal values – even while women remain restricted to the domestic sphere – and that this qualifies as a 'matriarchy'.

~~~~~~~~~~~~~~~

Ryan and Jethá (p. 133) try to encourage male readers to favour female autonomy and authority with the promise that societies where this is

---

[18] *Matriarchal Values and World Peace: The Case of the Minangkabau*, Peggy Reeves Sanday, 2nd World Congress on Matriarchal Studies
http://www.second-congress-matriarchal-studies.com/sanday.html

the case are "plenty sexy". If the evidence from the Minangkabau and the Mosuo is anything to go by this is a false promise. The Minangkabau are certainly not "plenty sexy": sex appears to figure quite low in their priorities and marriages are arranged in line with family alliances – there are no multiple sexual relationships here, apart from some polygyny. In the Mosuo, perhaps some younger men may get to have an increased number of brief sexual relationships with a greater number of these promised "girls" but the sum total of sexual activity looks to be far less than most Western men get, or men across societies generally access through marriage.

Ryan and Jethá (p. 134) paint a picture of most Mosuo men lounging around, relaxed and happy but the *Lugu Lake Mosuo Cultural Development Association* explain that the practice of having trading caravans has effectively ceased with the result that one of the primary male roles has been rendered irrelevant. So, they explain, *some* men can be found to be lounging around while women work hard but this is because of the need for a new male role within modern Mosuo culture. So it's not the result of a 'sexutopian' fantasy 'matriarchy' where men are only used for sex after all.

The bonobos are, erroneously, mentioned yet again as the poster species for a fantasy matriarchy. In terms of 'patriarchy' and 'matriarchy' what would we call a bonobo system? Bonobo females cannot sustain relationships with, nor pass anything to, daughters because daughters leave for good at puberty, never to be seen again, while mothers focus on their sons who stay with them for life. Bonobo sons can gain their status from their mothers and can even get their access to fertile females through their mothers. *Mothers act in the interests of their sons* and any influence of maternal status can only last for each mother-son generation and no further – she'll never know who her grandchildren are.

In some human societies mothers also lose their sexually mature daughters to another family and focus all their attention on their sons, supporting them in their male dominant societies. If we had the bonobo system where daughters leave at puberty and mothers focus on their sons, would we call it a matriarchy? Would the (imaginary) opposite system of sons leaving at sexual maturity and fathers acting in the interests of daughters for each single generation be a patriarchy?

We cannot use the bonobo system as an example of a matriarchy.

*

- Hrdy's arguments about the importance of shared parenting do not include any suggestion that men would provide any paternal care outside of monogamous relationships; she does not consider men to have made reliable parents at all.

- Ryan and Jethá present a false picture of foundling hospitals. Their implication that communal parenting would have allowed these offspring to survive will be contradicted by their upcoming argument for a deliberate disposal by our ancestors of any babies above the replacement level of an average of two offspring per woman.

- Canela virginity, rather than being irrelevant, is so important that for both sexes masturbation is strictly forbidden.

- Warao men are not free to have sex with whomever they like but may make a "skin payment" to a man in exchange for sex with his wife and to create an alliance between the two men.

- Pirahã women are trading sex for goods rather than simply 'sharing' sex with outsiders.

- Information from numerous more tribes from the Aché to the !Kung is again distorted to create a far more casual and sexy image of these peoples than exists in reality.

- The casual nature of Mosuo sexual activity is exaggerated by the authors and an image of female autonomy and "matriarchy" is falsely presented as a "plenty sexy" carrot to male readers.

- Almost all the peoples presented by Ryan and Jethá are settled horticulturalists or are otherwise not representative of pre-agricultural ancestors.

- No explanation is offered as to why marriage exists at all.

CHAPTER SIX

Monogamy and Jealousy

Monogamy

No one in evolutionary biology argues that social monogamy goes hand in hand with exclusive sexual monogamy, and neither do the evolutionary psychologists whom Ryan and Jethá accuse of upholding the 'standard narrative'. What evolutionary biology would argue, though, is that in socially monogamous species there cannot be selection for acceptance of extra-pair matings by a reproductive partner.

As we have already noted, humans are mostly socially monogamous or polygynous across all societies, with a few exceptional cases of polyandry and one case, the Mosuo, where particular factors came together to make socially recognized marriage more of an exception. If our pre-agricultural ancestors were really living like bonobos we would not expect marriage and socially recognized pair-bonding to figure so prominently across the world so we know that it is not the result of agriculture. Ryan and Jethá simply present us with a bonobo-like ancestor up to 10,000 years ago and use the absence of life-time exclusive sexual monogamy today as the evidence for that ancestor, completely ignoring our obvious pair-bonding. At least the 'standard narrative' does not leave out this rather important piece of the puzzle.

Ryan and Jethá (p. 136) mock the research into monogamous species such as the prairie vole, yet they earlier used (though erroneously) something that had come from these studies, i.e., from research into the AVPR1A (vasopressin receptor) repetitive microsatellite. They mistakenly latched onto the initial ape comparisons as crucial genetic evidence pointing to a similarity between humans and bonobos but not chimpanzees, also contorting it into an oxytocin link. As we saw earlier, further studies proved that the same genetic region is also shared with gorillas and Central African chimpanzees. No link has been found between these regions and differences between primate social behaviour either. The RS3 334 genetic variation Ryan and Jethá discuss briefly in their notes, which may have some small association with bonding ability in men and with autism, is actually part of this AVPR1A repetitive microsatellite.

It is one thing for the authors to mock (p. 134) the candy-coated representation of animal reproductive behaviour such as that portrayed in the film *March of the Penguins* but quite another to mock scientific research into genes, hormones, and behaviours, using animals such as the prairie vole. And throughout *Sex at Dawn* Ryan and Jethá are certainly not averse to their own candy-coated versions of bonobo and human sexual and reproductive behaviour; conflict-free promiscuity is no less candy-coated than is conflict-free monogamy.

The main point that the authors want to make is that monogamy, whether in the prairie vole or the many socially monogamous bird species, does not include sexual exclusivity. True. Neither does it *exclude* sexual exclusivity for most of the pairs.

Ryan and Jethá (p. 136) use an unreferenced quote from Thomas Insel, (Director of the National Institute of Mental Health (NIMH), and former director of Yerkes Primate Center) about the prairie vole, where Insel says: "They'll sleep [sic] with anyone but they'll only sit by their partners".

This unreferenced quote (and its conversational style with the use of "sleep" suggests it was a casual comment and said for effect) enables them to end their CHAPTER NINE (p. 137) with the line: "With such all-encompassing interpretations of the concepts [of marriage, monogamy, and the nuclear family], even the prairie vole, who 'sleeps with anyone', would qualify".

So, does the prairie vole 'sleep with anyone'?

In experiments where access to mates is controlled (prevented or enabled) a male or a female paired prairie vole *will* copulate with an individual other than its bonded partner. C. Sue Carter and Lowell L. Getz (1993) write about experiments which have shown that a female will mate both with her familiar male and with an unfamiliar male, but most nonsexual contact is only with the familiar male. *Once mated, both the male and the female of the pair will become exceptionally aggressive towards unfamiliar members of their own sex.* So there we have it again: this is mate defence or mate guarding, and it clearly functions to prevent the extra-pair matings under natural conditions.

The review article Ryan and Jethá reference[19] covers work done by Insel and his colleagues, including Carter. As the article says:

"The average person probably thinks of monogamy as a sexually exclusive relationship. Biologists, however, define the word a little differently. The monogamous animal is one that spends most of its time with one mate but is not entirely faithful, points out Insel. Most monogamous animals will, on occasion, mate with a stranger, he says."

"On occasion" mating with a stranger does not have quite the same impact as "sleeps with anyone", does it.

Insel and his colleagues observed the mate-guarding behaviour of the prairie vole. After mating, the normally timid male prairie vole will attack any strange male that happens by the nest.

---

[19] http://findarticles.com/p/articles/mi_m1200/is_n22_v144/ai_14642472/

The review article ends with an answer to Ryan and Jethá's remark (p. 136) that: "it's striking to note the scraps of comfort contemporary scientists find in equating human sexual behavior with that of a ratlike prairie vole."

The article explains:

"To the layperson, the study of vole society may seem like a frivolous occupation. Such studies undoubtedly reveal the fascinating details of a vole's sex life, but so what?

"According to Carter and other neuro-scientists in the field, research on voles is uncovering important clues to how brain hormones influence complex social attachments. 'By studying animal behavior, we are beginning to see the emergence of patterns of hormone usage', she says. Furthermore, by charting the course of such brain hormones in the rodent world, scientists hope to find additional pieces to the puzzle of what makes humans tick."

Ryan and Jethá want us to believe that human pair-bonding with mate-guarding and jealousy only arose due to agriculture. This is clearly incorrect. Or – though it's impossible from what they write to really conclude this – perhaps they do mean that there was pair-bonding amongst our ancestors together with open extra-pair sex? If they do mean this they have offered no mention or explanation of how or why humans evolved pair-bonding when bonobos did not, or explained how pair-bonding could evolve alongside indifference to extra-pair sex. They do note (p. 137) "the socially approved special relationship that often exists between men and women" but they give no explanation for such special relationships in our species. They are at pains to identify sexual activity outside of the pair bond but seem blind to the need for an explanation for the very existence of those pair bonds.

Against monogamy Ryan and Jethá are using evidence that pair-bonding across species can occasionally include extra-pair sex but so does the 'standard narrative', and the 'standard narrative' includes that rather significant factor that the extra-pair sex is not open but is carried out secretly. Ryan and Jethá want us to believe that ancestral promiscuity was open and accepted and even necessary yet there is no evidence from any species or from evolutionary biology that pair-bonding can exist alongside such promiscuity. Open promiscuity means no pair-bonding.

Perhaps the 'promise' of sex – whenever and with whomever – has blinded those who accept the *Sex at Dawn* argument to this problem? It wouldn't be the first time sexual arousal and sexual fantasy addled the brain.

Social monogamy does not mean sexual monogamy though it often can, and the absence of exclusive sexual monogamy does not equal sexual promiscuity (multiple open simultaneous sexual relationships). Sexual monogamy can result from successful mate-guarding or the lack of willing or desirable mates or because evolved and environmental factors means sexual monogamy leads to greater reproductive fitness.

The authors seem to think that monogamy depends on a complete lack of sexual interest outside of the pair bond but this is not how or why monogamy evolves. As we have already briefly considered in the context of Cai Hua's conclusions about the Mosuo, a desire for multiple sexual partners *and* a desire to prevent the socially monogamous reproductive partner mating outside of the pair bond are not mutually exclusive desires. This discordance is what Ryan and Jethá want to wish away.

In *The Battle of the Sexes Revisited* Helena Cronin (2006) asks us to picture a pair of monogamous South American titi monkeys:

"...in close embrace, their tails entwined, in sleep cuddled together, when awake always close, preferring one another's company above that of all others – and with so little sex difference that they look more like twins than spouses."

The male carries the infant in its first two weeks, and their parental division of labour reflects, she says, cooperation between their 'parental-investment' genes.

"But continue to trace the interests of their respective genes and we find that what might appear to be the very epitome of their happy pact – the entwined tails the constant cuddles – reflects instead genetic conflict. For they are mate-guarding: an adaptation that reflects an evolutionary legacy of less than perfect monogamy."

Mate-guarding would be unnecessary if the mate was 'naturally' sexually monogamous. If neither of the pair is 'naturally' sexually monogamous then why not both simply enjoy their sexual desires for other mates? The male is protecting his parental investment in his own genes when the female may benefit from mating with a superior male. She is protecting the male parental investment she needs when he could benefit from a greater focus on matings with other females. Cronin continues:

"So their conflict is over mate choice. And it is engendered by the very resource, parental investment, that their cooperation has created. What joins them together has also – among their genes – put them asunder."

Genes need the cooperation from the other sex to get into the future yet the differences between the two sexes means they also have conflicting reproductive agendas.

Birds

By now it should come as no surprise that some members of the many bird species that are socially monogamous are not necessarily sexually monogamous. Ryan and Jethá (p. 136) mention swans and bluebirds, and they say that most socially monogamous bird species have been shown to have chicks sired by extra-pair males. They think that this shows that male parental care is not dependent on 'paternity certainty', as if those males know that other males have sired those chicks and they really don't care. But, of course, if the males were really indifferent to paternity they would not be so concerned about mate-guarding, and the extra-pair copulations would not be so secretive.

In bluebirds (and in many waterfowl) the females will also lay eggs (known as egg dumping or nest parasitism) in the nests of other females if they can. In this species the males guard the females from other males, and the females guard their nests from this nest parasitism by other females (Gowaty, Plissner, and Williams 1989). This is interesting because it shows both the male and the female are potentially parenting unrelated offspring and it shows how selection therefore works on both mothers and fathers to avoid this. We probably don't really question the role of selection on these mother birds to avoid parenting the chicks of other females but the same applies to males: selection cannot act in the direction of indifference towards the parentage of offspring that are provided with parental care whether the parent is female or male.

Another very important point about birds is that they are often producing a *clutch* of eggs and the resident male will have fathered some, most, or – most often – all of those chicks within the brood. There is one species, the dunnock, where some females do openly mate with two males. In a situation not unlike that of the Aché and other partible paternity societies, the female dunnock will trade sex for male parental resources but she will not do so with more than two males. This is because the males adjust how much effort they put into feeding chicks in relation to their likely share of paternity. If there are more than two males as potential fathers of the chicks, any one male's chance of paternity is so reduced that it is not worth investing in the chicks (Birkhead 2000).

As well as males protecting their own investment in their own offspring females also show behaviours for protecting the investment that comes from their male. In the great reed warbler a secondary female will destroy the eggs of the primary female (Hansson, Bensch, and Hasselquist 1997). In European starlings the aggression from the resident female is enough to prevent a secondary female from breeding with the same male (Sandell and Smith 1996, Sandell 1998).

In house sparrows the conflict between females over parental investment from the same male leads to infanticide by the secondary female (Veiga 1990). Also in house sparrows, if the male takes over a female's nest after her fertile phase has ended he will destroy the young she produces so that he can ensure his own paternity of the eggs she later lays after mating with him. The females have evolved 'strategic ovulation retardation' as a counter to this risk which means that they are able to hold back the production of eggs for a number of days if their male is replaced by another one (Veiga 1993).

Direct punishment of females by males is not commonly seen though it has been investigated in shrikes (Valera, Hoi, and Kristin 1997). Mate-guarding is the common male bird behaviour and this is sometimes actually solicited by the female when she seeks protection against harassment from other males while away from the nest. White-fronted bee-eater females can be chased by many males and sometimes they are forced to the ground and mounted by up to six of them. To counter this, females call their male before leaving the nest; one study found that only 7% of escorted females were chased by other males compared to 70% of unescorted females (Emlen and Wrege 1986).

Similar findings have been observed in other species such as the common murre, barn swallows, waterfowl, cattle egrets, and great tits (Mesnick 1997). This is an important aspect of mate-guarding by males that is often overlooked in the excitement to show female promiscuity: females often can, and do, benefit from the protection one male can provide against unwanted sexual harassment. It is this protective mating alliance which may be the main factor leading to monogamy. Direct male parental investment is more likely a secondary outcome of monogamy because in socially monogamous mammals direct male parental investment is often not present.

Sarah Mesnick has considered the costs and benefits to females of these alliances with a male – costs such as reduced opportunity for free mate choice, and the need to be sexually available to the pair-bonded male lest he goes looking elsewhere – and she concludes that when sexual aggression constrains the daily activities of females a protective alliance may represent the "best of a bad situation" for them (Mesnick 1997).

We could just look at the extra-pair paternity of some chicks in socially monogamous birds and, like Ryan and Jethá, imagine that this shows just an 'enjoyment' of sex and not much paternal – or maternal – concern about parental investment but we would, of course, be wrong. What it shows is what we have seen already: the desire for multiple mates, especially in males who will use force if they can, and the desire that a reproductive partner does not have any extra-pair sex.

Across and within bird species there is a lot of variation in how much extra-pair paternity there is of the chicks that are raised together. In most cases the chicks are those of the two parents: one study, for example, showed that for blue tits there were no extra-pair offspring in 30 out of 48 nests, and in great tits none in 40 out of 55 nests (Krokene, Rigstad, Dale, and Lifjeld 1998).

As well as both sexes showing behaviours that limit the extra-pair copulations or extra-pair investment by their partner it should also be noted that for females not all of their extra-pair sex is solicited. If females cannot benefit from their male's mate-guarding behaviour and therefore avoid sexual harassment, the acceptance of unwanted copulations is what is known as *convenience polyandry*. Male sexual behaviour tends to be relatively straightforward in that males are expected, in most cases, to always potentially benefit from copulation. Female sexual behaviour is quite different with costs as well as benefits to be considered, so when females do accept matings it cannot be automatically presumed, as it is with the male, that the copulation *per se* is the benefit; it can be about avoiding the potential costs of resistance.

The existence of female sexual strategies, such as female sexual proceptivity to acquire food from males or to avoid infanticide by males, or sexual receptivity to avoid the costs of resistance, means that sexual behaviour of females cannot be assumed to mirror that of males. While the recognition of polyandrous mating by females across many species (including humans) is welcomed by most of us who recognize that females are not 'naturally' sexually monogamous, we must not forget that this does not make females the same as males.

Female mating behaviour is under very different selection pressures from male mating behaviour. Females often still want to say "no", and in this respect, choosing to attach themselves to one male so that they will not be sexually harassed by multiple males can be a sound reproductive strategy. Mutually successful mate-guarding can have its benefits and when it leads to parenting by two committed parents it can also benefit offspring and become an advantageous reproductive strategy for both sexes.

Jealousy

Ryan and Jethá suggest that our hunter-gatherer ancestors would have forced or obliged their members to 'share' sex with pretty much every one of the opposite sex in the group to keep the members bonded. Though they say this shared sex is widespread human behaviour the only 'evidence' they can come up with is from the South American partible paternity tribes, and even here we have seen that things are far more complex and far less sexually 'generous' than the authors would like us to believe. These peoples are also recent arrivals in South America, mostly horticulturalists, and their history involves a lot of tribal hostility and warfare, so they cannot be taken as the representatives of our hunter-gatherer ancestors who lived in Africa for most of the last 200,000 years.

In hunter-gatherers we often find groups with fluid membership mostly comprising mobile nuclear family units that are linked through marriage into a much larger network. Our ancestors evolved in Africa and there is no reason to believe they lived as modern horticultural tribes do today in the Amazon. Looking at one actual hunter-gatherer culture again, the Hadza, (referred to by Ryan and Jethá but not, unsurprisingly, in this context) Marlowe (2004b) writes that serial monogamy is the best way to describe the mating system though perhaps 20% of Hadza stay married to the same person their whole life. Divorce is often due to the pursuit of extramarital affairs. He writes:

"When I asked, "What happens if someone finds out his or her spouse has had an affair?" 38% of men and women said the man would try to kill the other man, 26% said a woman would fight with the other woman, 20% said a man would leave his wife, and 13% said the woman would leave her husband."

Hadza women value fidelity of a spouse as much as the men do, and Hadza divorce often occurs when angry wives leave their philandering husbands. Married women who have affairs run the risk of being hit or killed by their husbands so clearly, Marlowe writes, Hadza men care greatly about fidelity. He concludes:

"Hadza men want a good-looking, hard-working, fertile, faithful wife and Hadza women want a husband who is a good forager, good looking, intelligent, and faithful. Both sexes value character greatly, which probably reflects their desire for a stable, trustworthy partner. The Hadza are merely one society along the foraging spectrum, but their preferences are likely focused on traits which most impact reproductive success in a niche something like that occupied by humans prior to agriculture and are therefore a valuable case for comparison with the many studies of [the mate preferences of] college students."

Another hunter-gatherer group often referred to in connection to our ancestral past are the !Kung. From *Nisa: The Life and Words of a !Kung Woman,* Shostak (1990) writes:

"Sex is also recognized as tapping some of the most intense and potentially explosive of human emotions – especially where extramarital attractions are concerned. In such cases, sex is considered outright dangerous: many affairs that become known lead to violence, which, in the past, sometimes resulted in death. Except for those who intend to goad their spouses, therefore, people who participate in such relationships are extremely careful and discreet."

She continues:

"The best insurance against complications arising from love affairs is not to be found out."

And:

"Partly because of the lack of privacy in !Kung life, actual extramarital sexual encounters seem to be infrequent."

Ryan and Jethá do not refer directly to !Kung female sexual behaviour in *Sex at Dawn* but Ryan does refer to it in an unpublished interview he posted on his *Psychology Today* blog[20]. In this interview he states:

"...and Nisa says having one lover would never be enough because one makes you laugh, one brings you meat, another smells great...she has this sense of how each relationship fulfils a different part of her life."

Really?

Nisa actually says:

"One man can give you very little. One man gives you only one kind of food to eat. But when you have lovers, one brings you something and another brings you something else. One comes at night with meat, another with money, another with beads. Your husband also does those things and gives them to you."

"If [a woman] goes somewhere to visit alone, then [a lover] will give her meat, [another] will give her other food... Even if she goes with her husband, she should still have lovers. Because each one gives her something. She gathers from one man one thing, from another man something else, and from another, yet something else. It is as though her genitals were worth money - Pounds! Rands! Shillings (laughs)! She collects her gatherings from each different place until she has filled her kaross with beads and pubic aprons with money."

It is as though her genitals were worth money!
This is very different from the tale told by Ryan.

---

[20] *Ill-Fated Interview Part VI.* Published on August 31, 2011 by Christopher Ryan in *Sex at Dawn:* http://www.psychologytoday.com/blog/sex-dawn/201108/ill-fated-interview-part-vi   Retrieved December 28 2011.

Nisa also says:

"I like having lovers but their ways are to ruin my heart and to spill semen all over me."

Of a fight with the wife of one of her lovers she says:

"I pulled off my bracelets and hit her, hit her in the stomach, and she fell down. She got up and came at me, but I hit her again and again she fell. When she got up and came at me again I called her name and said: "This time I'll kill you". I started to laugh again, "I'll kill you, so you'd better go and sit down. If you come back I'll hit you so hard that it will leave you dead... I continued to yell at her, then jumped up and bit her hand... I said: "I'd like to beat you until you shit! ...After that he and I were no longer lovers."

And when it comes to any impression that hunter-gatherer sex is an 'anything goes' type of sex:

"Only rarely, if a woman is really frustrated, will she touch herself. Adult men also touch themselves, either in the bush or sometimes even in their huts. But only if they are refused by women. Women don't take men's genitals into their mouths nor do men kiss women's genitals. Men only kiss women's mouths."

There is a significant amount of conflict between men and women over sex and infidelity in the book *Nisa*. There is violence, including Nisa's father against her mother: he hits her in the stomach in one incident when she is pregnant. He also threatens to kill her over her infidelity.

Nisa herself has an unusual number of husbands and lovers, and of her four children none survived, which could be related to her unstable marital relationships and sexual behaviour. Her sexual behaviour is even criticized by a fellow !Kung as being like that of a man.

And, of course, it is the men who provide the meat; the sexual division of labour is as clear here as in all other societies. Again there is the interdependence of husband and wife, the nuclear family group makes up the basic unit, spouses make connections through in-laws enabling movement to different resources, and wider social networks depend upon marriage.

The Aka

We can also look at the African Aka hunter-gatherers' attitudes towards sex. The Aka are one of the most sexually egalitarian tribes: men and women often hunt together, and there is more direct father care of infants than in any other culture. Most are married monogamously but there is also some polygyny (Hewlett and Hewlett 2010).

Between the ages of 18 and 45 married Aka have sex 2 to 3 times a week and average sex three times on each of those nights. Though they do, of course, enjoy sex it is viewed more as work rather than pleasure and it is primarily about "searching for a child".

Oral sex and anal sex are unknown or very rare. There are no terms for masturbation or homosexuality, and the indications are that they too do not occur. Masturbation also appears to be rare in tribes in other areas as it has been difficult to explain to the men how to provide semen samples for fertility studies: in spite of lengthy and explicit instructions the semen samples they provide are still mixed with vaginal secretions.

The main reason for sex is procreation, and other noted reasons are for pleasure and to show love for a spouse. Interestingly, a husband says that showing this love for a wife is important for him as it means he will have something to eat because the woman the man has intercourse with will prepare food for him. This is an important point because we tend to think more about food resources the wife gets from the husband rather than the husband from the wife but the sexual division of labour creates an interdependence on which both spouses depend; upsetting either spouse could lead to the significant costs of withdrawal of their labour.

The younger Aka under 25 years of age were the most adamant about having sex to build a child. Hewlett and Hewlett contrast this with the Western cultural emphasis on recreational sex which "has also led some researchers to suggest that human sexuality is similar to bonobo apes… The bonobo view may apply to Euro-Americans, but from an Aka or Nganda [neighbouring farmers] viewpoint sex is linked to reproduction and building a family." They caution against the thinking that Western patterns of sexual behaviour – frequency of sex, reasons for sex, masturbation, and homosexuality – are common, if not universal and natural, when in fact cross-cultural data suggest the patterns in the West are quite unusual. Representations of human sexuality in many college textbooks "often reflect the interests and priorities of middle-class Euro-American cultural models".

Ryan and Jethá (p. 143) also note the unreliability of university undergraduates and modern Westerners for information on human sexuality and ask: "How can we expect to identify 'human universals' without including at least a few foragers", yet foragers are notably absent from their own 'evidence' of human sexuality.

Hewlett and Hewlett also note: "the commonalities shared by married couples in most cultures of the world: passion and love of mates, desire for privacy during sexual activities and jealousy over having intercourse with others."

Hewlett and Hewlett (2008) write that divorce is common amongst the Aka and in 64% of divorces the cause was the spouse sleeping with, searching for, or finding another mate. Marital violence is rare but hitting or slapping in response to the fear of losing a spouse is a common theme among the Aka. The Aka are mobile, immediate return foragers, dependent on extended family members of both spouses because they need to be flexible in response to food sources. Intra- and intergroup hostilities are infrequent and the nature of their violence that does occur is about protecting the commitment between spouses.

Living in nuclear families within extended family groups means that the Aka couple are not dependent just on each other for social and emotional support. But in the nuclear family unit within the extended family there is a strong interdependence arising from the sexual division of labour, and sexual fidelity is very important.

Wherever we look sexual jealousy does exist.

Ryan and Jethá (p. 141) make the point that if it is a question of genes then a man or women should be less concerned if their spouse had a sexual relationship with a sibling than with a stranger which they point out is not the case in the modern Western world. When we look at tribes that have polygynous marriages these are more harmonious when the wives are sisters: there is less conflict between the wives and greater care towards these children related as nieces and nephews. When wives are not related there can be serious rivalry between them; in the Dogon, for example, there is a higher mortality of children (especially sons) in the polygynous as opposed to monogamous marriages, and the mothers claim their sons are being poisoned by co-wives (Hrdy 1999a).

Polyandrous marriage is far less common and can be fraternal where the inherited land of the brothers is not divided and 'sharing' one wife keeps it intact down through generations. If brothers have few resources or there are few available women then sharing one wife can also be better than having no wife.

If we are just talking about sex rather than marriage then of course people in the modern Western world are going to be more distraught if, from all the many potential sexual partners, their spouse has a sexual relationship with a sibling rather than a stranger, with its destruction of trust and respect in two directions. Sibling rivalry can be intense – fatal, as we have seen, in some species – because sometimes the closer two individuals are the more likely they will be competing for the same resources. In the modern Western world a sibling acting to 'steal' a spouse when so many other potential mates are available can be felt as a particularly nasty outcome of sibling rivalry.

Where there is the tolerance of discreet sexual liaisons in the South American tribes we have looked at, these are people with a very limited number of potential sexual partners *and* they are very insular groups with intra-group marriage and breeding in many cases. If potential sexual partners, including spouses, are limited to cross-cousins or something similar then each individual has exactly the same small number of potential sexual partners as their same-sex sibling. It is interesting that, in spite of the potential need for a secondary father due to a high male mortality rate – which should not even matter if all the group cared for all the children – and in spite of the close relatedness of potential fathers, and in spite of the knowledge that closely related males might be lovers of a wife, there is still the jealousy, the competition between brothers, and the absence of open sexual liaisons.

When there are marriage arrangements between differently located groups or tribes these can lead to conflict and warfare as well as friendly links between the groups. The Mae Enga of Papua New Guinea have a saying: "we marry our enemies", due to the fact that women move between groups on marriage but warfare is also common between the men in these patrilocal groups (not unlike chimpanzees and, to a less violent degree, bonobos) (Bowden 1992).

The anthropologist Max Gluckman (1963) writes that the tribes he studied in East Africa also have a similar proverb: "They are our enemies. We marry them.'' The men marry the daughters and sisters of their actual enemies which also means that the wives are potentially the enemy within.

The South American Tukanoan people, as we have seen when we looked at Chernela's study of the Wanano, have in-coming wives who are very much "alien" or "other", reinforced by the fact that they speak a different language even though they are cross-cousins. Few potential spouses exist so there is the ambivalent nature of the relationship between the patrilineal brothers: they accept (by not thinking or talking about it) the discreet sexual liaisons between a wife and brothers and nephews but there is also competition over this limited number of females. Sexual relations between a wife and a man from another patriclan are an intolerable breach of the intact patriline and strictly forbidden (Chernela 2002).

Our ancestors over the last 200,000 years may have sometimes exchanged mates with distant tribes or tribal groups and had good relations with them due to these alliances, or may sometimes have had serious disagreements because of them. They may sometimes have become more isolated as small groups with restricted numbers of potential mates and spouses leading to more in-group marriages and more in-breeding within what might have been more of an extended family than a wide-ranging tribe. But the evidence remains for reproductive pair bond alliances, a strong dislike of sexual infidelity, plenty of sexual competition and sexual jealousy, and females exchanging sex for resources, i.e., not a promiscuous breeding system.

Zero-Sum Sex?

Ryan and Jethá (p. 141) argue that sexual monogamy is an expression of a "free-market" vision of human mating. They define monogamy as "individual male 'ownership' of female reproductive capacity". These few paragraphs in the book are at the same time particularly confused and particularly revealing: they are the most explicit paragraphs that connect private ownership of the 'means of reproduction' with private ownership of the 'means of production' and are therefore the likely crux of the authors' arguments.

If this were the 19th century and if it were polygyny that was being argued against then the authors could possibly be forgiven: some men 'owning' a number of women and many men left without any could fit the 'zero-sum', 'I win, you lose' argument between men. Monogamy, though, which is the more equal sharing of the women, is an attempted solution to this 'zero-sum' battling between men, not the cause of it. And when we have both sexes making their own mate choice then we have the complementary 'ownership' by women of the husband's reproductive resources.

Ryan and Jethá say that the "free-market vision of human mating hinges on the assumption that sexual monogamy is intrinsic to human nature". This argument is clearly wrong on many fronts: men obviously want the use of the reproductive capacity of more than one woman – they are naturally polygynous – and that's what creates the 'I win, you lose' outcome, not monogamy which is more about reducing the monopolization of all women by some men.

The authors' attempt at a solution is to combine polygyny with male-equality by enabling all men to have sexual access to all (or all attractive/fertile) females and therefore, rather than sharing women by having one wife each, they share by all having an equal pop at all desirable women!

What about the women? Later Ryan and Jethá will argue that women are built to accommodate this solution of being shared, and ancestral women 'naturally' would have had no reason to discriminate between men.

The authors discuss (p. 144) studies of sexual jealousy and the 'disconnect' between what women *say* they feel about a partner's sexual compared to emotional infidelity, and what the measurement of the women's *physiological responses* show. This is a hint of what will be followed up later with regard to a 'disconnect' between what women think their sexual arousal is and what their physiological responses say it is. We'll come back to this later but here we might note that a 'disconnect' between what women *think* or *say* they feel and what their bodies are supposedly saying (used by the authors' to argue that the body's responses are the genuine, ancestral, 'natural' ones) is an already familiar rape-defence argument (she said "no" but her body/eyes/behaviour said "yes").

What we have in these few paragraphs is the absence of the recognition of monogamy as the socially enforced sharing of women between men to reduce male-male within-group competition rather than an 'I win, you lose' situation. Of course most men (in the service of 'selfish genes' in their sperm) are not going to be content with sexual monogamy. Is all men sharing all women a better solution? Will we really find that women, if assured of the resources they need for their offspring, will accommodate this 'selfish gene' strategy in men to copulate with as many different (fertile) females as possible?

Have Ryan and Jethá really stumbled across the answer to what women want?

They think (p. 142) that our "extravagant sexual capacity", our "ubiquitous adultery", and the "rampant promiscuity" in both chimpanzees and bonobos is all the proof we need. The desire for a faithful mate and the sexual rejection of men by women are, apparently, social constructs.

What are Ryan and Jethá really imagining for our ancestors in terms of 'sharing sex'? Are we to believe that the removal of the need to access resources in exchange for sex makes all men desirable in the eyes of women? Does the sexual availability of all women make the women all equally desirable in the eyes of the men? Or is this fantasy of 'shared sex' really a male fantasy of not being sexually rejected by any of the sexiest females, especially for the want of some resources those females might find 'sexy' – that is, reproductively advantageous – in a male?

As we will see later, the authors' argument for sperm competition adds to this (implicit) argument for the removal of pre-copulatory female mate choice; in stark contrast they provide multiple references to the naturalness of male preferences for particular 'sexy' – that is, reproductively advantageous – female traits.

Sauce for the goose, anyone?

As long as a hierarchy of attractiveness exists in males *or* females and as long as people have similar preferences for similar traits in the other sex there will be competition. These preferences are connected in both sexes to reproductive fitness benefits. It would appear, according to Ryan and Jethá, that only female and not male sexual preferences are social constructs.

Fortunately for Ryan and Jethá's argument, women don't have to be sexually aroused for them to be 'shared' by all the men, but what about the men? Should they be punished for not being able to pleasure *any* and *all* females regardless of age or attractiveness? Is the removal of the female hierarchy of attractiveness as easy as the removal of that of the male? Is equality of desirability going to apply to both sexes or is this another argument where everyone is equal it's just that some (i.e., men) are more equal than others (i.e., women)?

If this *Sex at Dawn* story, on the other hand, is just about there being a number of simultaneous, mutually chosen sexual pairings within the group and female as well as male sexual preferences are in play, how do we then get around the competition and jealousies that the inevitable sexual rejection creates? How do we really combine sexual mate choice by both sexes and equal sharing of sex? By the social enforcement of 'sharing' – and, presumably, plenty of Viagra for men when it is their turn with the least attractive women? Should any social enforcement really be necessary if this 'sharing' argument is based on it being our natural, evolved sexuality?

All too often this 'sharing' of sex sounds like just another male argument over the distribution of women as if arguing about the equal distribution of food or other non-human resource. Unlike food, sex is something which involves the participation of another living being. What a male and a female bring to this interaction is different, and not least amongst that difference is the fact that DNA is making a one-way journey and there is no two-way sharing; the reproductive consequences of sex are very different for the two sexes.

Getting, or at least desiring, sexual access to attractive (i.e. fertile) females is the strongest drive most men (like males of other species) experience in life; getting round the problem of sexual rejection is strongly selected for, and men's fantasies are full of imaginary places where sexual rejection does not happen.

The shiny, superficially egalitarian wrapping of 'shared sex' in *Sex at Dawn* makes it no less of a male fantasy.

The Siriono

Ryan and Jethá turn to a South American tribe again to show us just how different jealousy can be. We met the Siriono earlier when Ryan and Jethá mistakenly represented them as having group marriage between a group of brothers and a group of sisters; now they say (p. 145) that the Siriono do not feel jealousy due to the spouse having lovers but because he or she is devoting too much time to those lovers.

The Siriono, they correctly tell us, use the same expression *secubi* ("I like") for everything they enjoy, whether food, jewellery, or a sexual partner. Sex, like hunger, is a drive to be satisfied and, though there are certain ideals of erotic bliss (i.e., pubescent girls, though Ryan and Jethá choose to omit this), any port in a storm will do.

Information about the Siriono comes from research by Allan Holmberg (1969) in the middle of the last century. This is another small South American tribe with many similarities to other tribes already described. The Siriono are in small groups (the two Holmberg discusses were 58 and 94 in number) who live together in the same house and have matrilocal, cross-cousin marriages: a male simply moves his hammock from near to his own parents to one near to the parents of his wife.

These people are near-starving and always hungry (Holmberg's reason for his studies was his interest in the connections between constant hunger and culture) with many quarrels over food and its allotment, hoarding, and the reluctance to share. The nuclear family is the fundamental social and economic unit, there is a sexual division of labour, and women are subservient to men. Most marriages are monogamous but the higher status better hunter males have more than one wife. If the wives are sisters, as they most often are in the polygynous marriages, there is less jealousy than when not. The wife with whom the husband has the most sex gets the most meat so the wives frequently vie with each other for his sexual favours.

In quarrels over sex, Holmberg says that men channel their aggression towards adulterous wives, and women channel theirs towards the women who have caused their husband to err; women are therefore believed to be the cause of most sexual disputes.

Sex is usually in the privacy of the bush and is a rapid and violent affair with few preliminaries: no kissing but there can be scratching and biting. If a man is alone with a woman in the forest he may roughly throw her to the ground and have sex without a word being spoken. Once a girl has gone through puberty rites, which can happen before she has actually reached puberty, she can be married; any sex with her cannot be seen as rape because 'rape' can only be applied to sex with a girl who has not gone through the puberty rites.

Food is the best means men have to obtain extra-marital sex, using game to seduce a woman who otherwise might not yield to his demands. An example is given of this. One man, Aciba-eoko, had tried several times to seduce one of his potential wives (socially legitimate extramarital sex partner, i.e., not too closely related). This woman had refused because she did not want to provoke a quarrel with her husband. One day the woman saw Aciba-eoko returning from the hunt with a fat peccary (a pig-like animal) and she was eager to get some of the meat. He was able to get her to agree to his demand for sex first and both entered, and later left, the privacy of the forest separately.

Ryan and Jethá imply there is an easy sharing of food and, because the sex drive like hunger needs to be satisfied, therefore an easy sharing of sex. In fact, the Siriono experience great anxiety over and obsession about food because there is never enough, and there is much quarrelling over its distribution. Once again we have women exchanging sex for meat and the discreet nature of this sex. And once again the greater detail of male-female relations shows something quite different from the relaxed and casual sex portrayed throughout *Sex at Dawn*.

The Canela again

Ryan and Jethá (p. 146) attribute a quote (unreferenced) to William Crocker stating that he is convinced that Canela husbands are not jealous. These are actually Sarah Hrdy's own words in her book *Mother Nature* (1999a), presumably her own *impression* from reading Crocker's *The Canela: Bonding Through Kinship, Ritual, and Sex* (1994), which she references. Ryan and Jethá have mistakenly thought these were Crocker's words.

So, what does William Crocker actually say?

As we have already seen, the Canela 'wives' who engage in most of the extra-marital sex are actually young teenage girls yet to become mothers. A girl becomes married to the first man she has sex with when she is perhaps 12 years old. She then has her 'sex work' to do, motivating the men to work – and in the past to bond for warfare – but once she becomes a mother her work becomes that of childcare and domestic duties. So girls in this sexually 'free' state are only technically married, the girl can be 'married' to a succession of men with whom she is currently having sex, and this period ends with the conception of a child. The man she is with at that time is her husband for life (Crocker and Crocker 1994).

One occasion of sexual jealousy by a young husband was recorded by Crocker. During a sing-dancing Great Day celebration a young woman was in a line with the men so her husband knew that most of the men in the line would be having sex with her. Consumed by jealousy he grabbed his wife and dragged her inside and stood at the door waving his machete. This was ultimately to no avail because several days later, while he was away, the men took his wife to the woods, "passing her from shoulder to shoulder like a racing log". Her mother followed with a machete but could not catch up to them and the men took turns with the young wife.

Women who cooperate receive gifts from the men, but if the husband shows jealousy his wife will receive nothing. There are clearly many factors involved in the compliance of women with the desires of the husband *or* with those of the other men, and sex in exchange for resources rears its head yet again.

Elsewhere Crocker writes (Crocker 2002):

"Sex jealousy is thought by many anthropologists to be culturally determined. If this were the case, in a society like the Canela in which extramarital sex was not only condoned but required, sex jealousy would not exist. However, sex jealousy did exist and had to be suppressed even before outside influences began to interfere with the extramarital sex practices. Thus sex jealousy could not be culturally determined, but may be determined at some psychological or psycho-physiological level."

So Crocker does not say that he is convinced that Canela men do not feel sexual jealousy. As noted earlier, when the use of sexual access to young teenage girls to control young men became unworkable because they were no longer using those bonds for warfare, the young men then increasingly openly expressed their sexual jealousy and 'possession' of their wives.

So is jealousy natural?

Ryan and Jethá (pp. 147-148) tell us that fear is natural and jealousy is an expression of fear. Whether or not someone else's sex life provokes fear, they say, "depends on how sex is defined in a given society, relationship, and individual's personality".

The authors compare it to the jealousy siblings might feel over a mother's love and say: "Why is it so easy to believe that a mother's love isn't a zero-sum proposition, but that sexual love is a finite resource?"

Actually, a mother's 'love' is not always so abundant that there's enough for everyone – Hrdy in *Mother Nature* shows us, as we saw earlier, that human mothers are quite unusual amongst primates in making strategic decisions over whether to invest in any particular child or not. When more than one offspring is seeking parental investment at the same time, whether age-mates or, as in humans, of different ages, then sibling jealousy and rivalry, and maternal neglect or the disposal of one or more of the offspring so the other or others can prosper, are not outside of our nature. Parental investment is a finite resource over which siblings do compete and often with good reason.

But reproductive relationships between two parents are something quite different because these are joint ventures with an unrelated other – or less closely related than parent-offspring or siblings – whose behaviour is crucial to the survival or extinction of one's gene copies. The reproductive gateway for genes to make it into the future *is* a finite resource. Other animals evolved behaviours in response to this finite resource without any conscious need to know about numbers of genes copies and genetic relatedness. We too might never think in terms of these gene copies but evolution has taken care of this without any need for (and probably does better without) our conscious awareness of what is going on.

Ryan and Jethá ask what it might be like if women had no dependence on a particular man for their own and their children's well-being. What if that fear is gone? They think this was the case with our ancestors but clearly their evidence so far is seriously flawed. But could it be the case for our descendants? What if we were like the Mosuo, they ask.

As we have seen, in the Mosuo the women engage in less sex than in other cultures. For reasonably attractive 15-30 year olds there can be sexual relationships with a number of partners, though women with young children tend to be less involved. Older people – and that includes women over 50 years old – are not acceptably sexually active, and people past 30 or 40 tend to stick with one partner. So it's not that much different from us, and when women have easier access to resources they seem to be less, not more, available for sex.

We could also note that Mosuo women having sex over the age of fifty being frowned upon does not sit well with what the authors say (p. 143) about the difference between the sexuality of twenty-year-old women and women three decades older: "Most would agree that a woman's sexuality changes considerably throughout adulthood – for the better, if conditions allow". The conditions of the Mosuo, it appears, do not allow.

Ryan and Jethá (p. 149) write: "The anachronistic presumption that women have *always* bartered their sexual favors to individual men in return for help with childcare, food, protection, and the rest of it collapses with the many societies where women feel no need to negotiate such deals". Yet in every society they have brought up, women exchanging sex for resources exists. Even in the Mosuo, where the need is least of all, there is still some gift-giving by the men – and a lot less sex from the women.

Ryan and Jethá again use a quote from E. O. Wilson (1978) writing that sex is primarily a bonding device and only secondarily for procreation. But, as pointed out previously when these quotes from Wilson have been used, he is not writing about sex-as-bonding in the way they are – his criticism is of the religious, particularly Catholic, attitudes towards sex as only being allowed for procreation, and he is writing of non-reproductive sex as a bonding device between a couple *in* a monogamous relationship.

E. O. Wilson, in the very pages from which the authors select their quote, writes about the effects of the differences between being a sperm producer and being an egg producer, and the sexual conflict between the sexes in most species. He notes the great variation across cultures but that the flexibility is not endless and beneath it all there lies the general features that conform to evolutionary theory. He argues that frequent human sexual behaviour is to strengthen the pair bond between a man and a woman.

The use of sex for bonding between a man and a woman in a monogamous relationship can be argued as being a mechanism to cement the reproductive compromise each accepts. The intensity of sex and emotions in our monogamous relationships across cultures is certainly not to be found in the bonobos or any other species. Neither is the degree of empathy and compassion we can feel, and it is hard not to see the likely relationship between all these human traits and the needs of our human offspring. Perhaps women have experienced stronger selection for these traits because they are traits which have evolved in connection to the needs of our offspring; maybe men only have them as a by-product, or selection has been weaker in men because of the ability of our male ancestors to sometimes reproduce successfully without caring for mother or child.

When both sexes want to possess their partner while still desiring other partners, monogamy can only be a compromise – but one that most likely benefited our offspring in the environment in which we evolved. Some degree of open polygyny and secretive polyandry was no doubt normal. 'Sharing' women by reducing the incidence of polygyny so that most men could at least have one wife would have been an important step in our sharing and cooperation and mutual respect of the needs and feelings of others, including children. Couples no doubt could and did separate and enter new marriages, and people no doubt discreetly engaged in sex on occasion outside the pair bond; if obviously or too often divorce was the likely result. The sexual division of labour and the interdependence of spouses also would have acted against the willingness to jeopardise marital relationships.

Ryan and Jethá talk about something called 'sex' which can be shared as if it has *nothing* to do with reproduction, and as if it comes without a history of naturally and sexually selected traits for increasing the spread of some genes over others. Evolutionary theory plus the evidence from across human cultures tells us something quite different. Only in the modern Western world has 'recreational' sex not only become separated from reproduction but it has taken precedence over reproduction. In most species males focus more on mating effort, females more on parenting effort. Our species needed parental input from the males too but perhaps we should not be surprised if this is a relatively weak or variable trait in men, nor be surprised if opportunities for returning to focusing just on sex and leaving out reproduction – and concerns about children – are grabbed with both male hands.

Sex at Dawn, rather than a plausible potential explanation of our evolution, increasingly reveals itself as a contemporary middle-class, child-free, sex-obsessed, male fantasy projected back onto prehistory. It may increasingly become our present but it certainly isn't our human past. And 'recreational sex' is not what creates the future.

*

- Conflict-free promiscuity is no less candy-coated than conflict-free monogamy.

- Mate-guarding behaviours and the secretive nature of extra-pair sex in monogamous species are ignored.

- Ryan and Jethá ignore the evidence for sexual jealousy and the importance of sexual fidelity in hunter-gatherers.

- The misrepresentation elsewhere by Christopher Ryan of what *Nisa* says about her lovers is a revealing distortion.

- The 'singles scene', childless women, and the emphasis on recreational sex rather than children are modern Western phenomena.

- Ryan and Jethá wrongly equate monogamy with 'zero-sum', 'free-market' private ownership of the 'means of reproduction'.

- Women are presented as a passive resource, and the only argument is over how this resource should be distributed amongst men.

- Holmberg states that the nuclear family is the fundamental social and economic unit in the Siriono.

- Siriono sex is not of the casual nature suggested by Ryan and Jethá. Rather than an easy access to sex akin to an easy access to food, the Siriono live in constant hunger and there are quarrels over food as well as sex, and food is the best means men have to obtain extra-marital sex.

- Crocker does not say that he is convinced that Canela men do not feel sexual jealousy. He does say that he believes sexual jealousy is not culturally determined.

CHAPTER SEVEN

The Way We Were?

Part III of *Sex at Dawn* is where Ryan and Jethá present their picture of the day-to-day social world of our pre-agricultural ancestors. To sum up this picture, over a period of about 200,000 years our ancestors lived in groups of about 50-150, were widely dispersed, and they constantly moved to new food sources as the local food resources dwindled. Such abundant resources meant there was no cause for competition within the group or between different groups who would have had infrequent contact anyway. The constant walking and the desire for leisure meant that our ancestors never got enough food to build up fat reserves which was especially important in stopping the women translating fat reserves into too many babies. If more than a replacement number of babies were born the group not only killed the less than healthy but the surplus healthy ones too.

Even, they argue, during periods of adversity, perhaps due to climate changes caused by natural events such as volcanic eruptions, no advantage could be had by any individuals within these groups. And, of course, we must not forget the casual sexual promiscuity acting as the social glue, presumably including sex between paternal brothers and sisters, and even between fathers and daughters. While other species have members of one sex emigrating and immigrating at puberty, Ryan and Jethá do not include anything like this in their picture of our ancestors, and by treating the small, extended family group as a single bounded entity through time it makes, as it did with Morgan's mistaken imaginings, inbreeding and incest an unavoidable part of their picture.

All for one and one for all.

The original affluent society

Ryan and Jethá wish to present a view of our ancestors in direct opposition to the "solitary, poor, nasty, brutish, and short" one of Hobbes which they see as being currently dominant. What we get is one incorrect version of our ancestors replaced by another.

It is most likely true that our hunter-gatherer ancestors did not generally lead lives that were solitary, poor, nasty, brutish, and short. The idea that their lives were the opposite of this and that they lived in the "original affluent society" was one that was first put forward by Marshall Sahlins in the 1960s. This argument was primarily a reaction against the idea of 'progress' and the improvement of human societies over time. In proposing peaceful and contented hunter-gatherers, Sahlins wanted to counter the image of them as people who had lost out to the agriculturalists and had suffered as a consequence (Kelly 2007).

This 'affluence' was an overstatement to grab attention, not unlike the headline-grabbing exaggeration of the 'peace and love' aspects of bonobos has been a reaction against the focus on the less attractive aspects of chimpanzee behaviour. Ryan and Jethá have used these exaggerated, counter-images of ancestors and bonobos and embroidered them with further exaggerations and misrepresentations of human sexuality. Unfortunately, like 'Chinese whispers', or the way myths are created and grow, an element of truth at root grows into increasingly distorted stories as the news spreads amongst those only too willing to believe and propagate confrontational or sexy headlines.

Writing about the origins of the "original affluent society" theory in his book *The Foraging Spectrum*, Robert Kelly (2007) writes that attention became centred on Sahlins' claim that hunter-gatherers do not work a lot, and in so doing anthropology "replaced one facile stereotype with another".

One problem Kelly notes is in the definition of the word 'work'. Only time spent acquiring food in the bush was counted as work, not the time spent processing food, making tools, childcare, carrying water, collecting firewood or cleaning. More accurate estimates of time spent foraging and processing food demonstrates that, while there is variation across hunter-gatherer societies, some hunter-gatherers spend seven, eight, or even more hours a day 'working'.

When it comes to sharing food this is often 'demand sharing' where the person with the food is badgered by others, and some group members still manage to hide what they have in order to avoid sharing. Women restrict their own foraging so they do not have to face demands to share from people outside of their own family. Hunters will eat meat, or the high calorie fat, where they have made the kill, so they help themselves first before returning to share with others. How meat is then distributed is not necessarily equal, and women and children often receive a lot less than do the men.

Egalitarianism, the sexual division of labour, and the family

Kelly writes that egalitarianism does not mean that all members have the same amount of goods, food, prestige, or authority. It does not mean that everyone is equal in what they have but that they have equal access to food sources, to the technology needed to acquire resources, and to the paths leading to prestige. What is important, as Ryan and Jethá also argue, is individual autonomy. In reality, though, some members have higher status and greater access to resources than others, particularly notable with regard to good hunters, and especially pronounced between men and women. With regard to men and women, the more time they are apart in their respective roles and the more hostile the environment, the more men tend to view women as subservient.

In their book *The Evolution of Human Societies,* Johnson and Earle (2000) write that the idea of 'the original affluent society' downplays the problems that are posed by seasonal and periodic hardships. Under changing conditions aggregation and dispersion of small groups occurs, and under these conditions the family as the basic unit group is an effective way to live: with about five to eight members the family is the basic subsistence unit. When resources are highly localized about five families will join together for a period while at other times larger numbers of families may come together temporarily.

This flexibility provides the mutual benefits of larger aggregations, such as the arrangement of marriages and other social activities, as well as providing for the exploitation of resources. When resources dwindle and tensions increase the families can again disperse elsewhere. Individuals have their own networks of kin, including in-laws, which gives them numerous connections to other families and camps. The family unit is the smallest unit that is self-sufficient because of its sexual division of labour and the strong interdependency of the wife and husband.

Johnson and Earle also write of the personal hostilities that can arise, especially between men. These are often over women and can lead to homicides which are fairly common. Organized warfare, they say, is not common. The evidence is strong that fighting between men over women within the group is a common occurrence and that hostilities between groups are often over women. As well as being due to sexual desire for the same female, a man is also concerned that he may lose a wife he depends on economically. Competition and conflict can also arise over arranged marriages because of the importance of these arrangements both to the individuals and to the families concerned due to the crucial social networks and connections marriage opens up.

Sharing *is* important but it also creates tensions between individuals who are providing for themselves and their families and are also under pressure from others to share what they have. The mobility of people is even a result of this tension: family units move to particular resources and can group together for a period of time but then resources dwindle and tensions build so people move on again. It is the connections through marriage that have created the multiple links to multiple resources and made this fission/fusion possible.

Hunter-gatherers are 'fiercely egalitarian' in that egalitarianism has to be worked at: 'upstarts' actively need to be stopped from taking authority over others. Boehm (1999) writes in *Hierarchy in the Forest*, that those who seek authority (and need to be stopped) are usually males, and egalitarianism pertains more to males than to females, though females are not as powerless as in more modern times. Inequalities in authority between men and women vary markedly across hunter-gatherer societies, with small to large advantages usually going to males. Inequality and the reactions against it are most likely related to male competition for females which Knauft (1991) writes is also the leading cause of hunter-gatherer homicide.

In the book *Nisa,* Shostok (1990) writes:
"Despite the prominence of !Kung women men generally have the edge. One reflection of their dominance is the pressure they can exert on their wives to accept other women as co-wives in marriage...he gains recognition and status in the community, and he extends his social and political influence to include his new in-laws, their village, and their foraging grounds... Many women become furious when their husbands suggest it...fights between co-wives, even between sisters, are fairly common."

So here we have the greater authority of men in a relatively egalitarian hunter-gatherer society and the importance of marriage in the creation of social, political, and economic networks.

All societies also have a sexual division of labour. The sexual division of labour is highly significant in human evolution and it is clearly connected to the constraints human babies place on their mothers, and the role of hunting by males.

The clearest picture we have of hunter-gatherer living is one where the nuclear family is the basic economic unit with temporary social and economic rewards bringing the family units together in larger groupings. As localized resources dwindle, and social tensions increase, these groups then disperse again. Marriage, with its sexual division of labour and the interdependence of husbands and wives, is what creates the web of kin relations and connections that enable access to wider information on the availability of resources.

Though there is cooperation and sharing in the extended family groupings of foragers there are also shortages that do affect all these peoples and will have done so in our past. It is naïve to think otherwise. For example, Sarah Hrdy writes about reverence for the elderly and how this varies across different cultures. Old women have lower status where hunting is particularly important, such as the Eskimos or the Aché, and higher status where they are important foragers and providers such as the !Kung and Hadza. She suggests that the hard work of these ageing foragers may be so that they stay useful to others due to a fear of being left behind – or worse.

Hunting meat is very important for the Aché, and foraging contributions by old women are of much less importance. In the Aché, as among the Eskimo, euthanasia is practiced. Hrdy writes about a startling interview with an old hunter in which he reminisces to a time when just the sound of his footsteps on the leaves of the forest floor struck terror in the hearts of old women. He was the socially sanctioned specialist in eliminating old women deemed no longer useful: coming up behind an old woman he would strike her on the head with his axe (Hrdy 1999a).

Holmberg (1969) also writes that in the Siriono the sick and the elderly are simply abandoned when they cannot travel with the group. He writes about one sick middle-aged woman whom the band all left behind without even a word of farewell. Her remains were discovered by Holmberg about three weeks later at a camp a little distance away from where she had been left; she had tried her best to follow the band. As Holmberg says, her failure to keep up with the band meant she experienced "the same fate that is accorded all Siriono whose days of utility are over."

Population

Ryan and Jethá give the impression that our ancestors *intentionally* kept their population in check by deliberately keeping themselves undernourished so that the women were too underweight to conceive and, if that failed, through infanticide. They accept that child mortality was high, as is the case in extant hunter-gatherers where child mortality can be 50% or more, but they prefer (p. 203) deliberate infanticide as the explanation for most of this child mortality in our past rather than it being a consequence of seasonally poor resources and disease. But as Kelly (2007) says, while mobility may act against the spread of contagious diseases it does, on the other hand, lead to contact with new strains of parasites found in different regions and can produce a chronic state of poor health.

Deliberate infanticide can only account for the wilful killing of newborns or young infants but a high child mortality rate includes children dying beyond these early weeks so this suggests that undernourishment, injury, infection, and disease are likely causes of those deaths. Though they obviously did not have the diseases that have arisen since agriculture there would still have been human variants of the infections and infestations that affect all species simply because pathogens are very much part of the natural world, and these would have contributed to early death especially in combination with undernourishment. (For some discussion of this see Hill, Hurtado, and Walker 2006, Gurven and Kaplan 2007).

Studies of hunter-gatherer mortality rates show that about 20% of Hadza infants die in their first year, and nearly half of all children do not make it to 15 years of age (Finkel 2009). For pre-contact, forest living Aché only 42% survived to age 50 (Hill *et al.*2001).There is variation in the mortality rates in adulthood, for example, the Hiwi of Venezuela are hunter-gatherers with a high adult mortality rate of about 2.3% per year compared to about 1.1 to 1.3% for pre-contact Aché, Hadza, and !Kung. Only 51% of Hiwi 10-year-olds in the pre-contact period were expected to reach age 40, compared to 72-76% of all pre-contact Aché, Hadza, and !Kung. In the Hiwi 36% of all adult pre-contact deaths were due to warfare and homicide. Most adult killings were due to either competition over women, reprisals by jealous husbands (on both their wives and their wives' lovers), or reprisals for past killings (Hill, Hurtado, and Walker 2006).

Ryan and Jethá (p. 208) make a point that severe caloric restriction is the only thing that has been shown to prolong life in mice and potentially in other mammals. They argue (p. 209) for the leisure benefits our ancestors gained from eating just enough to forestall serious hunger pangs, including the benefit of the time available to play with children, though how women on such severe calorie-restricted diets managed to actually produce and nurse these children is not explained.

What is even more of a contradiction is that the authors later (p. 259) write of "the amazing power" of female breasts and of curvaceous women, and that this is an indication of millennia of promiscuity. They mention the Venus figurines, dating back tens of thousands of years, which are hardly (see image, p. 281, this volume) the bodies of calorie-restricted women! They can't have it both ways.

We also noted earlier that the Siriono who live in that very condition of hunger actually live in a state of constant obsession about food, with quarrels and accusations of hoarding and refusing to share, and the men using food to seduce the otherwise reluctant women. It is near-impossible to believe that ancestral males would not use food in exchange for sex, and that the extra food obtained by women in this way would not be translated into offspring. Women who accepted additional meat or others foodstuffs in exchange for sex would have out-reproduced those who 'chose' to stay hungry and infertile and presumably much less curvaceously sexy.

Ryan and Jethá tell us (p. 206) about a study of the Waorani of Ecuador which showed that they were free of most diseases and had no evidence of health problems such as hypertension, heart disease, or cancer (Larrick *et al.*1979).What they don't add is that Larrick and his colleagues found that 42% of all population losses were actually caused by Waorani killing other Waorani. In a subsequent study Beckerman *et al.* (2009) report:

"At the time of first peaceful contact, the Waorani...were the only human inhabitants of a region about the size of New Jersey, located east of the Andes between the Napo and Curaray rivers. Although occasionally traversing the region or hunting in it, neighboring groups were afraid to settle there, and with good reason. When the Waorani found invaders, they speared them. Their reputation for ferocity was earned by violence against each other as well as outsiders. In a genealogy of 551 individuals going back over 5 generations, Larrick et al. found that 42% of all population losses were caused by Waorani killing other Waorani. These homicides accounted for 54% of male and 39% of female deaths at all ages. In addition to this 42%, another 8% of Waorani were killed by neighboring cultural groups... The abductions of women and children by other cultural groups (about 9% of individuals lost to the Wao population), and their flight to escape from homicidal pursuers (about 5% of losses), brings to 64% the proportion of all Wao population losses attributable to warfare and the threat of violence. This figure refers to the entire population, including women, children, and infants, as well as adult males. Most outmigrants, both captured and voluntary, died of infectious diseases shortly after leaving their territory, so the majority of losses due to flight and abduction should properly be classified as mortality rather than migration."

So that is *sixty-four per cent* of a small population which was spread over a vast area dying due to in-group violence and warfare. Is the absence of 'modern' diseases really the only thing we needed to know about the Waorani? Apart from anything else it is certainly interesting that the 36% who lived amongst but survived this violence do not develop the stress-related illnesses we might expect.

With regard to modern day stress, Ryan and Jethá (p. 210) ask why, if natural, are we still so vulnerable to stress. While not denying that modern day stress and modern lives are very different from what our ancestors experienced, this statement suggests again that the authors have a fundamentally mistaken understanding of natural selection, as if natural selection acts in terms of individual well-being rather than in terms of differential reproductive success. Other species suffer the negative consequences of stress, especially males in sexual competition. Testosterone has its part to play in this, and males of many species suffer loss of condition and infestations and disease during breeding seasons. Remember the antechinus? Sexual competition for him means he literally drops dead from stress. Do we ask why natural selection has not acted against this in the antechinus? If we do then we don't yet understand natural selection.

~~~~~~~~~~~~~~~

Ryan and Jethá particularly want to counter the argument by Malthus that populations inevitably expand at a greater rate than resources which inevitably leads to competition over those resources and starvation for some. Of course the population increases that came with agriculture greatly increased competition over land and resources but it is no more reasonable to argue that the forager predecessors of these settlers over hundreds of thousands of years were *intentionally* choosing a low population growth than it is to argue that the orangutans intentionally choose to keep *their* population numbers so low. Did our ancestors really limit women to an average of two surviving offspring each and simply kill those that were born to any woman beyond this number?

Ryan and Jethá confuse birth rate with population growth, thinking that the former has to be high to produce the latter. They make the point that we didn't breed like rabbits but they miss the point that rabbits can breed like rabbits and still have no population growth. For the population to grow it only needs for there to be an average of more than two surviving offspring per female; four surviving (and reproducing) offspring per female and the population can double every generation but even only a little above two and the population can grow quite rapidly over time.

Presumably the authors do not think that our ancestors from the time of our common ancestor with the chimpanzees and bonobos some six million years ago intentionally kept *their* population numbers down, so these ancestors would have been faced with the same 'competitive' natural selection and differential reproductive success as other species. Only, according to Ryan and Jethá, once our ancestors had, *due to* that natural selection, evolved into modern humans with the traits that could only be *a consequence* of that natural selection, did we then *choose* to snub our noses at natural selection and go hungry.

There's a long way to go between the common ancestor we share with chimpanzees and bonobos some six million years ago and the one that eventually became the modern human some 200,000 years ago, and further still to the one that spread out from Africa some 70,000 years ago. There are a few points we must note, though. Firstly, it is now believed that we started to become more bipedal very early in our split from the chimpanzee line when we were still living in woods or forests (see, for example, White *et al* 2009, who also suggest that our common ancestor with the chimpanzee was quite different from the modern chimpanzee). This would have presented a very early problem for infant carrying if the infant could not travel on the back of a mother who was no longer primarily using quadrupedal locomotion (Amaral 2008).

Losing gripping toes as feet evolved over time for walking would have made it even more impossible to hold onto an upright mother, and then with the loss of fur, at least by 1.2 million years ago if not well before that time, there was nothing to grip anyway (for fur loss see Rogers, Iltis, and Wooding 2004). How infants and young children would have been carried by bipedal mothers during those hundreds of thousands of years before we were able to invent a way to strap an infant to its mother, and the impact of needing to be held by a mother's arms, is a question which has not yet been satisfactorily answered.

The changes over the last one or two million years coincide with our ancestors moving away from the shade of the trees and evolving traits in a new open environment which now probably included extensive scavenging or hunting on the savanna and the sexual division of labour. Along with the increase in brain size, increase in body size, decrease in sexual size dimorphism, the novel female reproductive fat stores, assisted childbirth, and all the additional effects from producing large-brained and highly dependent offspring, this is the most likely time and place for the strongest selection for pair-bonding in our species.

This is the time when our ancestors were *Homo erectus,* some of whom moved out of Africa and into Europe and across Asia to become Neanderthals and similar *Homo* species that would be replaced by the more recent movement of *Homo sapiens* out of Africa from about 70,000 years ago. For comparison it is also the time, around one million years ago, when a group of our chimpanzee cousins became separated south of the River Congo and evolved into the bonobos.

A lot happened after our split from the common ancestor we shared with the chimpanzees and bonobos. Movements of (now extinct) *Homo erectus* ancestors out of Africa half a million years or more ago suggests a difficulty in finding food sources from this time rather than a mere ambling along to the next food source in an environment which was, according to Ryan and Jethá, "chock-full" of food. Population numbers may never have been great in Africa yet at least some ancestors had long ago sought food sources beyond that continent.

Evolution is about change. It is influenced by climate, by travel to new environments to access food, by predators and prey, by competition for the same resources between species and within the same species, and by sexual competition. The changes our ancestors have gone through over millions of years: bipedalism, fur-loss, the increase in brain size, decrease in digestive tract, changes in our thermoregulatory system, changes in our teeth and jaws, our resource hungry and highly dependent offspring, the complexity of our tool-making and social systems – these changes do not arise in stable environments with stable food supplies. Novel genetic mutations not only have to arise but the bodies they are in have to out-survive and out-reproduce others *of the same species* who do not have those advantageous novel traits if they are to spread; that is why they are 'advantageous' and are 'selected'.

Ryan and Jethá (p. 160) argue that our ancestors lived in a world "chock-full" of food but chose to stay a bit hungry. The abundant sex would, presumably, have involved thin, barely-fertile women without much in the way of reproductive fat that is stored in the breasts and buttocks of human females. It would be surprising if some groups did not take advantage of those abundant resources to alleviate their hunger, plump out the females, and turn out more than two surviving offspring per female – and therefore out-reproduce those who chose to be hungry and who disposed of excess offspring. Even without any intention to increase their numbers, not being hungry plus the attractiveness of sexy (i.e., fertile) curvy females would be enough to lead to more than two surviving babies per woman. Unless conditions were not that great.

A regular use of infanticide in order to avoid having to go and access those abundant resources needed to feed children or their lactating mothers, and the 'casual sexual promiscuity' of the *Sex at Dawn* story, unfortunately brings to mind the bawdy *Barnacle Bill the Sailor* drinking song where the answer to the maiden's question: "What if I should have a child?" is: "We'll smother the bugger and fuck for another". Nice.

Ryan and Jethá propose that our ancestors, at least from 200,000 years ago, were living in a world of plenty where the resources of food and sex were in abundance but were choosing to stay hungry. After a couple of hundred thousand years of this 'Eden' we apparently forgot who we were, got our appetites back, and before we knew it we were drowning under babies we no longer chose to dispose of at birth.

An apparent slow population growth does not tell us, as Ryan and Jethá say it does, that our ancestors were deliberately balancing their numbers with the resources available to them. It does not tell us there was no competition, especially in times of shortages which inevitably occurred. It does not tell us that food and resources were equally shared, and especially it does not tell us that 'sex' and reproductive costs were shared.

A more realistic picture is one where we were much more like other species, satiating our hunger if possible and not living in a world of easy abundance. Most of our mortality would have been beyond our control but the lack of resources would, as Hrdy argues, have sometimes led to acts of infanticide by mothers rather than, as Ryan and Jethá suggest, infanticide being a fairly easy, everyday mechanism to deliberately keep the population from growing, avoid the work of foraging, and enjoy plenty of leisure time.

Anthropologist Renee Pennington looked at demographic rates of hunter-gatherers and concluded that a slow population growth rate is not plausible. What she found to be more likely are periods of rapid growth and decline; boom times probably balanced by periods of epidemics and famines (Pennington 2001).

The most likely scenario for our ancestors is one of good times and bad times; population growth mixed with population decline. Variation between people in status, attractiveness, skills and abilities, and in strategies to obtain extra food (relatively more important for females) and extra matings (relatively more important for males) would have made a difference in reproductive success especially in those bad times.

Ryan and Jethá (p. 208) *also* write that there would be fasts and feasts which is at odds with their argument for a world chock-full of food and the deliberate choice to stay a little hungry. They also, as we'll see later, only accept and support variation in reproductive success between males (and never even consider variation in reproductive success between females) when it is played out in the context of sperm competition within the female's body. This is essential for their scenario of male equality: men are made equal by the removal of female pre-copulation sexual preferences and mate choice so that all men are equally successful in gaining access to the woman's body, even if they are then unequal within it.

Warfare

Warfare in our ancestors was probably quite rare. Relatively small groups widely dispersed may well have had infrequent contact and chosen avoidance over conflict. Only perhaps during seasonal or periodic shortages might there have been conflict over the limited resources that were then available.

Ryan and Jethá (pp. 183-185) criticize the use of evidence of tribal warfare from tribes in the Amazon, Papua New Guinea, and Australia. "Are these groups representative of our hunter-gatherer ancestors?" they ask (p. 185). *"Not even close"* they say (emphasis in original). Fair enough. But it is similar tribes from these regions they use as evidence to support their own argument for shared paternity and the lack of concern about paternity certainty in our ancestors, so the tribes *they* present to support their own argument are, by their own admission, *"not even close"* to being representative of hunter-gatherer ancestors.

Perhaps there is even some connection between tribal warfare and the greater willingness by men to share women?

We saw this most clearly in the Canela with the connection between the sharing of women sexually and the bonding between men for warfare. Warfare is certainly a common factor in many of the Amazonian tribes we looked at earlier. Ryan and Jethá also argue for the role the sharing of women can have in male-male bonding, including by modern day sports team members which can be viewed as a modern substitute for real, warrior 'brothers in arms'.

Bonobos do not have this male-male bonding so the relationship between male bonding for 'warfare' and the sharing of the same females by those males is very much more in tune with chimpanzees than bonobos. In fact, it is the lack of male-male alliances and bonding in bonobos, and the greater incidence of male-female connections, that enables the bonobo females to have higher status compared to chimpanzee females, and reduces the inter-community violence by males.

The (not so) mysterious disappearance of Margaret Power

Ryan and Jethá (p. 187) write about "the mysterious disappearance of Margaret Power" whose 1991 book, *The Egalitarians - Human and Chimpanzee: An Anthropological View of Social Organization,* argued that chimpanzee inter-group and intra-group aggression was due only to the provisioning of chimpanzees with bananas. Ryan and Jethá think it strange that this argument has 'disappeared'.

Part of the misunderstanding of chimpanzee relations that led to Power's argument was the result of the fission-fusion nature of their communities which initially made it appear that chimpanzees from different groups were showing friendly behaviours when they met. But now we know that this fission-fusion occurs *within a community* and that males do not have friendly relations with other communities while females move from their natal to their breeding community at puberty. So the friendly male interactions that were so often seen were those between males of the same natal community.

No doubt the provisioning of bananas and human encroachment has greatly affected the behaviour of some populations of chimpanzees – just like the provisioning of bonobos with sugar cane brings out an excessive amount of pseudo-sexual behaviour in them – but now we have evidence for similar inter-community aggression in non-provisioned chimpanzees which is why provisioning is no longer considered the cause of aggression between communities.

For example, Boesch and Boesch-Achermann study non-provisioned chimpanzees in the Taï forest. They write (Boesch and Boesch-Achermann 2000) that the strategies used by the Taï chimpanzees in territorial attacks are elaborate "and reflect a dynamic evaluation of the forces present and an anticipation of the consequences of their actions". Some of these faculties, they say, are similar to those used in hunting.

Territorial patrols involve at least four males and mostly include deep incursions into the territory of neighbours. Some may simply lead to loud drumming before retreating. In others, if enough males are present, there is attack and counter-attack. With more than eight adult males there is often a frontal attack, ideally a surprise attack, and bad bites can be suffered. If numbers are equal there can be back-and-forth attacks. Females can also provide support from the rear with noise and drumming.

Another strategy is the 'commando' attack where a group of males silently moves deep inside the neighbours' territory looking for strangers and chasing them. On two occasions an isolated mother was found and held for a time while being bitten and hit.

Boesch and Boesch-Achermann write that all long-term studies on chimpanzees have shown aggressive inter-community interactions and aggressive defence of territory. The only exception is Bossou where the community has little or no connection with other chimpanzees.

Actual fatalities of adult males caused by adult males from a neighbouring group are rare, but aggression and violence is not. There have now been reports of lethal aggression at Loango, Gabon (Boesch *et al.* 2007) and Taï (Boesch *et al.* 2008). The latter paper is a fascinating study of inter-community relations and the perspective of females, including both voluntary sexual interactions and the capture and temporary 'imprisonment' of females. Boesch *et al.* do refer to Power and her argument but note that lethal attacks have been observed throughout the range of chimpanzees in Africa which "strengthens the claim that this behavior is natural in chimpanzees, contrary to claims that humans induce it".

We can also note that there have now been observations of lethal aggression *within* communities too (Fawcett and Muhumuza 2000, Watts 2004).

So there is no 'mystery' about the disappearance of Margaret Power's argument.

In bonobos the females travel together more, and males and females associate together more in mixed-sex parties, so there are not the same conditions as in chimpanzees for joint male defence of the whole territory containing all the individual females spread out in their more isolated ranges. As we saw earlier, there is still the animosity which can include aggression and violence between males from different communities, and following an encounter with a neighbouring community more males have been seen to associate with the females where the encounter took place.

It took a number of years before inter-community aggression was observed in the chimpanzees; hunting by chimpanzees came as the first shock before their inter-community aggression added to it. The discovery that bonobos also hunt monkeys, among other species, does not make them inter-community aggressors on a par with chimpanzees but it may be a similar progression of discovery as in the chimpanzee which is why the hunting of monkeys by bonobos is a significant discovery.

In our ancestors the chimpanzee type of inter-community aggression was probably rare, but so was the bonobo-type because neither chimpanzee nor bonobo social systems allow for friendly interactions between males from different communities. Men are not likely to have remained for life knowing only the small number of males in their birth group while all other males remain forever strangers as happens in both chimpanzees and bonobos. The ancestral human males would not have been defending territories containing females who had wandered into their particular patrilocal territory at puberty and made it their reproductive home, raising their offspring alone.

With the evolution of our resource-hungry, long-term dependent offspring and the necessary sexual division of labour in humans, our pre-agriculture ancestors, like hunter-gatherers today, were more likely often living in small nuclear family units that came together in clusters of family groups when resources or social factors made it possible or necessary. Pair-bonding, not casual sex, made it possible for males to expand their social networks beyond their small birth community.

Ryan and Jethá (p. 190) admit that fighting over women was a possibility but they think that this claim would depend on population growth being important and that women could be traded like livestock. Along with this they say (p. 191) that "women and men would have been free to move among different bands in the fission-fusion system typical of hunter-gatherers, chimpanzees and bonobos". This, again, is a misunderstanding of a crucial difference between us and the chimpanzees and bonobos (also noted regarding Margaret Power above). Chimpanzee and bonobo males *cannot* move between communities, and females can, in most cases, only transfer when they have their sexual swelling 'passport' at puberty, i.e., females transfer from natal to breeding community when they reach sexual maturity, they *do not* merely move between different subgroups within their natal community, but leave it altogether.

Chimpanzees and bonobos *do* have what is called a fission-fusion system. The community of 20 to 100 members divides into separate foraging parties. Chimpanzee females, as we have seen, may spend most of their time alone with their offspring, but members of the community do travel together in various groupings when females have their sexual swellings and will be mating with males or when there are enough resources to forage and feed together or when males join together to patrol or hunt. Bonobo males appear to be alone more often than are chimpanzee males but bonobo females are usually with other females and are often also joined by males.

In both species the males remain in their birth community for life so they only know males born in that group, young females before they leave at puberty, and sexually mature female immigrants. The males of all communities are restricted for life to their birth community. The fission-fusion only happens *within* the community comprising natal males, immature natal females, and immigrant adult females. The significant contrast between humans and our chimpanzee and bonobo cousins is the absence in our ape cousins of peaceful interactions between neighbouring communities while in humans there are extended periods of friendly contacts – sexual, marital, and social – between 'communities' who will also at times become enemies.

Hunter-gatherer fission-fusion extends well beyond the community limits comparable to those of chimpanzees and bonobos. In chimpanzees and bonobos pubescent females visit neighbouring communities before making their choice of where to stay and mate and breed. At some point in our history this inbreeding-avoidance mechanism, common to most species, became of interest to family and wider group members rather than it being a more secretive and individual-choice matter for the female herself. When female transfer becomes a matter of group concern then arrangements and tensions between groups regarding the acquisition of their breeding females in exchange for their sisters and daughters would be expected to develop.

It is this in-breeding avoidance mechanism that opened up 'family' links between individuals in different 'communities' of humans.

Female chimpanzees and bonobos do not 'want' (due to selection against close inbreeding) to mate and reproduce with the males in their birth group who include a father, brothers, and other close relations. Neither, presumably, would our ancestral females if they once lived in this type of community. The chimpanzee and bonobo males who welcome pubescent females with sexual swellings into their community are not doing so because they are interested in population size; a male interest in sex with attractive new females is all that is needed. Presumably our male ancestors were no different.

Attracting or capturing fertile young females would be expected in a male philopatric species and this has nothing to do with any conscious desire to increase the population size in humans any more than in our ape cousins. The foundations for the movement of females between groups would have already been in place in our ancestors. Alliances between human groups would have included this exchange of fertile females between those groups – and disputes would also have erupted and females sometimes captured in raids on other groups. Ryan and Jethá have completely missed the deep roots of this movement of females when they say that fighting over women would have depended on population growth being important and that women could be traded like livestock.

So two important things to note are firstly that the chimpanzee and bonobo fission-fusion system is strictly limited to the single, small community which males cannot leave, and secondly, our ancestral social system meant the movement of pubescent females from their birth group to their breeding group would be influenced by people in both groups and not be a lone venture for the female concerned.

Ryan and Jethá (p. 171) write of ancestral group sizes of up to 150. This is the number of people with whom we seem to be able to maintain a reasonably close relationship. For Ryan and Jethá this has also been taken as the number of people who will be 'sharing sex' –

people who, they argue, have been known to each other for life. This implies a bounded, inbreeding group, and they relate it to the chimpanzee communities where females mate with all or most of the adult males within the group but they have not incorporated the emigration and the immigration of pubescent females. This movement of females creates the gene flow between groups and, in time, creates gene flow to more distant groups and therefore throughout the species. Without gene flow via this movement of females the inbreeding communities would be genetically isolated and eventually evolve into sub-species and then separate species.

Humans may have been dispersed and in relatively isolated populations for periods of time but consistently breeding only within separate and distinct groups of 150 is extremely unlikely. Individuals within any grouping will have had different and overlapping connections to other people which could be increased with any further connections created by pair bonds or marriage. With the increasing sophistication of language and the communication of information, the connections made possible through marital ties will have extended potential social networks to many hundreds at a tribal level.

If the argument in *Sex at Dawn* is that the sharing of sex is important to group cohesion, as, perhaps, in the bonobo model, then for humans to extend their 'cohesion' beyond the size of a bonobo or chimpanzee community of about sixty members implies that the females would be having to 'share sex' with increasing numbers of males and well beyond the numbers chimpanzee and bonobo females are faced with. Apart from the potential costs to females, especially the younger and more desirable females, of the excessive sexual requirements placed on them to stop the increasing numbers of men competing, we would also expect even greater sperm competition in our ancestors than exists in chimpanzees and bonobos.

What we need is a model that takes into account the probable initial free movement of females to a group of 'strangers'. We need to take into account the initial absence of any friendly interaction between males of neighbouring communities where these females have come from, no movement of males between communities, and we need to show how these male-male connections could have arisen. We need to understand how even groups of 150 are still part of a wider gene pool that defines a whole species and are not bounded, inbreeding, extended families in isolation, at least not for any significant length of time. It is through the pair bond and then socially recognized marriage that individuals become linked to *new* people as *'in-laws'*. Rather than a network of 150 people being the same 150 people for all members of that discrete group, each individual has a different though overlapping network of blood-kin and kin through marriage.

Pair bonds made it all possible

The picture Ryan and Jethá paint of our ancestors, right up to the beginnings of agriculture, is one of a bonobo-like system. They do not present any real explanation of how human social networks managed to evolve beyond the initial very small patrilocal groups where, if like chimpanzees and bonobos, no friendly links existed between males in their different natal groups. Their only answer is sex and more sex, with an increasing number of sexual partners over a wider network but this is then limited to an inbreeding, bounded population of about 150.

It is pointed out more than once by the authors that there are no other social primates with a sexually monogamous breeding system. That's true, but there are social primates with polygynous breeding units. Hamadryas baboons, geladas, mandrills and drills form one-male units within larger bands. These are believed to have developed from the more usual multimale-multifemale baboon system. Interestingly, savanna baboons will also subdivide into polygynous units during harsher conditions (Barton 1999).

The environments our ancestors lived in over millions of years after leaving the dense forests of our ape cousins would have been novel and harsh, and we know that many hominin cousins over this time went extinct. Polygynous units may well have been the most common reproductive and foraging units over much of that time with larger mixed groupings only as we evolved into *Homo*.

Ryan and Jethá do not deal with the millions of years between the time we shared a common ancestor with chimpanzees and bonobos and 200,000 years ago when we became fully modern humans. In truth, most of what happened through those years can only be conjecture, especially as it applies to social and sexual behaviours. But there is no justification for simply accepting that the modern bonobo (or the modern chimpanzee) is the correct mating system model for our species from 200,000 years ago until 10,000 years ago, while there is very good reason to acknowledge the evolution of multiple differences between humans and our ape cousins and that reproductive pair-bonding is one of those differences.

It is clear that hunter-gatherers, like all other human societies, all have marriage. Kin are recognized through fathers as well as mothers, and relationships extend not just to blood relatives but to those related through marriage. It is this extensive network of ties through in-laws as well as blood-kin that enables the mobility and the fission-fusion of family units with access to multiple resource locations.

If we look at a bonobo community, when a male is born in the group he has his relationship with his mother and in some cases this bond with his mother lasts for her life-time. In chimpanzees the males are more bonded as allied defenders of their territory. A daughter in both species will have her main bond with her mother and perhaps siblings that are still around her mother but when she reaches puberty she leaves and any ties with her natal community end. In encounters between two communities the reactions of the males towards each other are not friendly. For our ancestors to have developed connections between males from different groups there had to be bridges built between them. The transferring females are the obvious means by which this was achieved.

So how?

Bernard Chapais in *Primeval Kinship: How Pair-Bonding Gave Birth to Human Society* gives the most thorough description of the likely answer. In what follows I include the main points but would recommend reading his book for the most thorough understanding of the crucial nature of pair bonds and how promiscuous mating could not have given rise to the complex sociality we know as humans.

Either our very earliest ancestors were already polygynous and polygynous units amalgamated to form multi-family communities, or polygynous units developed within multimale-multifemale communities as we see in a number of baboon species. This was most likely due to a combination of male mate-guarding behaviour and female attachment to one male who provides protection from harassment from multiple males. This association with a particular male would not only give some protection against multiple males in her new community but also protection from the resident females who will not welcome a new immigrant female.

Chimpanzee females can face aggression from resident females when they join their new community and can sometimes receive protection from one or more males (Boesch and Boesch-Achermann 2000). Bonobo females have to ingratiate themselves with dominant females and their entry to their breeding community is also stressful. Male polygynous and mate-guarding preferences, incoming females benefitting from attachment to one male, and a fission-fusion environment where foraging females need to be with a male protector, when we put these together we have polygynous units within a multimale-multifemale system.

Offspring acquire their own social understanding from their mother's associations with others in the group, so a mother who associates primarily with a particular male would have offspring who grow up also associating with that male. The relationship between a mother and a particular male will also be a preferential mating relationship and the offspring will more likely be the offspring of that male. Male parental investment is not a factor at this time but the female benefits from association with one male in that her foraging is not constrained by multiple sexually aggressive males or by feeding dominance behaviour from multiple males and probably females too.

From only having a mother as a mediator of social relationships, as in bonobos and chimpanzees, the association between a mother and a particular male means her offspring also have that male as a mediator. This, if we were like other African apes, would be in the context of a wider community of related males. The mother will have joined the community as a 'stranger' from another community so she provides no maternal kin links, but being attached to a particular resident male creates kin relations through him for her offspring through his behaviour towards his own mother, the male his mother had a preferential relationship when he was young (probably his father), his brothers, and so on.

Daughters will leave at puberty which in chimpanzees and bonobos means the end of her relationships with members of her birth group. If that daughter forms a particular association with a particular male in her new breeding community she has preferential ties with that male and through him she has links with his closest kin. The transferring daughter now has preferential ties to a father and his male kin in her natal group and preferential ties to a pair-bonded mate in her new group. These are males who share an interest in her but *without* sexual competition over her: the males in her birth group are blood-kin while her preferred male in her reproductive group is her sexual partner. They have the potential to become allies. The link between males in the two groups has in effect been *revealed* by the act of pair-bonding, something promiscuous mating cannot do.

Over time in our evolution the constraints our offspring put on their mother's ability to be self-sufficient led to a sexual division of labour which added to the interdependency and made a male-female pairing the basic self-sufficient foraging unit. A fission-fusion foraging system would reduce to a male-female pairing as the most basic unit with further benefits from the joint hunting by men and the joint foraging or shared parenting by females in larger groupings. An increase in monogamy and a decrease in polygyny would result from increasing egalitarianism between males, and possibly competition between females not willing to share a male. And, of course, individual self-interest would make this a far from perfect system: sexual infidelity and potential sexual threats would always be close to the surface creating a heightened interest in the sexual behaviour of others.

When we looked at the South American tribes some of them stated that 'everybody's child is nobody's child'; in a similar way, everybody's kin is nobody's kin. The only known adult kin relationships in chimpanzees and bonobos are mothers and their adult sons, and sometimes maternal brothers. Only by revealing *specific* ties between people through fathers via pair bonds can kinship ties beyond those initial few come into play. Through these relationships the links, and therefore the potential alliances, between males from different groups can be revealed, and ultimately relations between 'stranger' males can be 'pacified'. Females have, Chapais argues, become the peacemakers through the development of specific pair bonds.

This is how extensive networks most likely developed in our forager ancestors and why marriage ties are so important, in fact crucial, in human societies. Ultimately males could move and reside with the 'wife's' kin rather than being limited to their own blood-kin and small natal group. An 'unknown paternity' system like that of chimpanzees or bonobos simply does not allow for this kind of establishment and extension of kinship and the pacification of relations between males from different birth groups, something which in time led to tribal associations for our ancestors.

Ryan and Jethá (p. 171) take their likely group size of 150 from Robin Dunbar's studies of the connection between the size of the neocortex and social group size (Dunbar 1992 and 1993). More recently Dunbar (2010) has given particular consideration to the evolution of pair-bonding in humans. This human group size of about 150 is, he says, about triple that of the chimpanzee but human groups are also more cohesive than chimpanzee communities, reconvening, for example, in overnight subgroup camps.

Dunbar notes the social stress female primates experience from male harassment (with possible threats of infanticide) and harassment from other females which has been shown to reduce female fertility. In larger groupings as we have in humans this harassment would be even greater so a female allying with a male minimizes social stress for females. Dunbar argues that pair-bonding evolved to counteract the social stress of larger groupings and he suggests that the increase in brain size after 500,000 years ago was therefore a consequence of pair bonds.

Whereas Ryan and Jethá imagine females simply accommodating the sexual interests of more and more males, Dunbar notes that this would not be possible beyond the size of a chimpanzee or bonobo community and that the protection of one male would have been necessary.

The initial polygynous and monogamous behaviour was most likely a result of females seeking a 'bodyguard' rather than for paternal provisioning which would come later, probably along with the sexual division of labour which created additional interdependence between the sexes. Females could initially simply benefit from their mate-guarding males as 'bodyguards', relieving them of some of the disadvantages of male dominance behaviours, sexual harassment from multiple males and sexual coercion, plus the stress faced by an immigrant female amongst established females.

The egalitarianism between males in the 'sharing' of females was not that they all had an equal 'pop' at the same females but that they each had a monopoly of one female, though higher status males would have been 'allowed' more than one wife. This does not mean exclusive sexual monogamy, of course. Men would have taken advantage of their own status to be openly polygynous or, if monogamously paired, have taken advantage of other casual sexual opportunities. Women would have used sex to acquire extra resources, and both sexes would have dissolved their 'marriages' and started new ones too. And, no doubt, as we see across human societies these extra-pair sexual relations would have caused tensions and upset and conflict, and even homicides as we see today. But the stable male-female pair bond created the means to increase group size and to open up extensive social, political, and economic networks; it was, and despite many problems has always been, crucial.

In our evolution it was not casual sex that created cohesion as group size increased but pair-bonding and ultimately marriage and the kinship networks that could be created through marriage. The increase in group size cannot occur in a chimpanzee or bonobo system because the breaking down of the barriers between communities in those species cannot happen without the pacification of the adult males; this required the multiple bridges that could only be revealed through a female and her particular association with one male.

It is a simplistic attitude to look at chimpanzees and bonobos, to look at ourselves as not being 'naturally' sexually monogamous, to look at the negatives of monogamous marriages, to forget much of our particular human evolution, and to then conclude that casual promiscuity must be our true 'nature'. But it simply does not fit with the unique complexity of the expanded networks of human societies. Casual promiscuity would have kept us in small, mutually intolerant, male philopatric groups.

*

- The 'original affluent society' was a term used to grab attention rather than being an accurate description of hunter-gatherer living.

- Hunter-gatherer men generally have higher status and greater access to resources than do women, and amongst the men the good hunters do better.

- The nuclear family is the basic hunter-gatherer subsistence unit.

- Men fighting over women is common and this not uncommonly leads to homicide.

- The 'fierce egalitarianism' of hunter-gatherers applies mostly to the men rather than to both sexes.

- Intentional caloric restriction as it is applied to women and their fertility does not fit with the male sexual preference for curvaceous female bodies, and especially not those depicted by the 'Venus' figurines.

- Ryan and Jethá write about a study of the Waorani of Ecuador to show how free they were from modern diseases, but they refrain from providing the interesting additional information that *64%* of these people – men, women, and children – died from violence, mostly amongst themselves.

- Rather than *intentionally* keeping population numbers stable, our ancestors most likely experienced periods of a more rapid population growth followed by periods of decline.

- All long-term studies on chimpanzees have shown aggressive inter-group interactions and aggressive defence of territory.

- The chimpanzee and bonobo fission-fusion system is strictly limited to the single, small community which males cannot leave and which females move between primarily only at puberty.

- Promiscuous mating could not have given rise to our complex sociality.

Increasing group size and social networks would have meant, following Ryan and Jethá's argument, females having to mate with increasing numbers of males and presumably adapting to far greater sperm competition than in chimpanzees and bonobos. The authors do not argue for this *greater* sperm competition but they do argue that there is evidence for similar sperm competition until ten thousand years ago and the arrival of agriculture. So now it is time to look at our bodies and see what they can tell us.

CHAPTER EIGHT

Body Talk

Sexual Dimorphism

In their Part IV Ryan and Jethá look to our bodies for evidence about our mating behaviour and (p. 213) consider it rated "XXX". They begin (p. 216) with body-size dimorphism and note that humans are similar to chimpanzees and bonobos in that the males are 10 to 20 per cent bigger and heavier than the females. Then they look at gorillas and note that the gorilla male is twice the size of the female: a sign of male competition over females and a polygynous breeding system. The gibbon, in contrast, is virtually sexually monomorphic and has a socially monogamous breeding system.

Conclusion: human males have not been fighting over females "in the past few million years". Our body-size dimorphism is similar to that of chimpanzees and bonobos who both have promiscuous mating systems so the authors' argument is that this is most likely to have been our mating system for the past few million years. In their introduction the authors (p. 11) define this "few million years" as being the time of *Homo erectus*, who actually existed from about 1.8 million years ago. Before that, they say, we had a polygynous breeding system.

As the authors note, studies of australopithecines suggest greater sexual dimorphism than is found in humans today. As well as *A. africanus* other hominin species, such as *P. robustus, P. boisei,* and *H. erectus,* are believed to have had greater sexual size dimorphism (Carnahan and Jensen-Seaman 2008). We don't know what the sexual dimorphism of the last common ancestor with the *Pan* lineage was but it seems reasonable to keep in mind that the *Pan/Homo* common ancestor and/or our hominin ancestors had a sexual size dimorphism greater than humans, chimpanzees, and bonobos today.

Our hominin lineage and that of the chimpanzee diverged into different species because they responded to different selection pressures in different environments, and we know our own lineage evolved many changes over those millions of years. Many of our hominin cousins over that time went extinct, and these are species we are more closely related to than we are to chimpanzees and bonobos. While the *Pan* and *Homo* lineages evolved to have multimale/multifemale social systems – though multifamily in the case of humans – these were two different evolutionary journeys: two different species experiencing very different selection pressures in very different environments and leading to two different outcomes.

We can again consider the fact that chimpanzee and bonobo males remain in their birth group. When males disperse from their birth group – as is normally the case across social species – they then have to try to enter a new group of 'strangers' before they can breed. Competition with established males can be fierce, such as in baboons, so we would expect that a large body size would be beneficial to males in social species with the more usual female philopatry. But if males are staying for life in their birth group and they cooperate in the defence of that group then this selection pressure on males is reduced and it would be one factor in the reduction of the sexual size dimorphism. Philopatric males do not have to experience the sometimes solitary and dangerous periods as bachelors followed by the struggle of entering a new group of 'stranger' males and the competition with these established males.

When it is the females who are the immigrants, as in chimpanzees and bonobos, they are the sex that faces competition with the unrelated, established females of the new group. When we have this male philopatry it is females who could therefore face some selection pressures for increase in size, and so the size difference between the sexes can be reduced from both directions: less in-group competition between males and more in-group competition between females.

It is critical to consider selection acting on both sexes, and to consider all the other environmental and reproductive factors that can influence evolution. In our own evolution we cannot simply ignore traits like the universal human sexual division of labour with its high interdependence of the sexes in small family units as we see in hunter-gatherers today. It would be very odd for human groups that have been untouched by settled living and agriculture to be somehow affected by these changes happening at a distance, which, though they never deal with this, would have to follow from the arguments of Ryan and Jethá.

The evolution of sexual size dimorphism is certainly not clear-cut. Another factor to consider is that as body size increases for the species so does the degree of sexual size dimorphism, and taking this into consideration the sexual size dimorphism of humans is strikingly low. We are relatively large-bodied compared to other primates, and simply due to species body size humans should show greater size dimorphism than we do, so compared to our relatively small-bodied cousins our sexual size dimorphism is actually on the low side (Martin, Willner, and Dettling 1994).

We also need to consider the selection pressures of our large-brained offspring on the size of mothers in the last million years or so which quite likely selected for bigger females while having no influence on the size of males. There are the reproductive fitness costs and benefits of how long a female body should keep growing before diverting energy into reproducing, and this affects what her optimal adult body size will be. For our ancestral mothers, growing that much bigger before reaching sexual maturity may have had a reproductive advantage. And, of course, the selected changes in the human female pelvis will also have impacted on overall human female body size.

We also cannot ignore that our ancestors made more use of tools and weapons so these means of competition or predator defence have also to be taken into account, especially with regard to males. Chapais (2008) argues that they would have equalized relations even more in our male-philopatric ancestors, so selection pressures for bigger males and bigger canine teeth would have been reduced.

And then we have another factor to consider. As David Puts (2010) points out, women are unique among primates in having copious reproductive fat stores. When only fat-free mass is considered men are 40% heavier, have 60% more total lean muscle mass, 80% greater arm muscle mass, and 50% more lower body muscle mass than women. The sex difference in upper-body muscle mass in humans turns out to be similar to the sex difference in fat-free mass in gorillas. These sex differences in muscularity translate into large differences in strength and speed, and men have about 90% greater upper-body strength and about 65% greater lower body strength. Using *these* comparisons we easily fall within the polygynous range.

So we need to take into consideration the weight of our unique human female reproductive fat stores – a major human sex difference – in our comparisons of the sexual size dimorphism of the apes. These fat stores are intrinsically connected to our unique reproductive needs which makes comparisons with other species far from simple. Our sexual size dimorphism and how it relates specifically to the human way of reproducing, with the sexual division of labour and the human male's role in providing for young via provisioning females, cannot be ignored.

The fat-free comparison gives a large difference between the human sexes which would correlate with polygyny. *On the other hand,* our sexual dimorphism could be explained as being due to a significant reproductive division of labour in a species that is essentially a monogamous breeder. The near-identical bodies and behaviours found in other monogamous species result from very similar selection pressures experienced by the two parents but selection for monogamy in humans can be argued as having produced two quite distinct but interdependent parental roles resulting in significant human sexual dimorphism.

With a little more recognition of the multiple factors involved in the subject we start to get some information that counters Ryan and Jethá's argument (p. 219) that if monogamous we should be the same size, like gibbons, and if polygynous men should be twice the size, like gorillas. Simply looking at sexual size dimorphism as it initially appears to be in humans and then placing us with chimpanzees and bonobos completely overlooks just how selection has acted differently on the human ape. We can use the generalizations about sexual size dimorphism as a starting point but certainly not as a conclusion.

Ryan and Jethá (p. 217) also argue against polygyny in our ancestors because "the cultural conditions necessary for some males to accumulate sufficient political power and wealth to support multiple wives and their children *simply did not exist before agriculture,*" (emphasis in original).

But we know that it is common for some men to be polygynous in hunter-gatherers, as we saw in the !Kung where the very reasons men seek more than one wife is to increase their social and economic network. Many of the South American tribes also have some polygynous marriages, such as the Siriono whom we looked at earlier. Ancestral females, like hunter-gatherer females, were not dependents in the way Ryan and Jethá suggest but hard-working foragers on whom husbands depended economically as well as for the social and political benefits that came with increased numbers of family connections. Polygyny certainly does exist in hunter-gatherers and it is incomprehensible how the authors could think otherwise.

Sperm Competition

Relatively large testes usually correlate with sperm competition which is selected for when ovulating females mate with more than one male. Ryan and Jethá (p. 220) argue that sexual competition between our ancestral males took place between sperm from multiple males within the body of the ancestral female rather than between the ancestral males themselves.

Of course, Ryan and Jethá have created our ancestors in the image of the exaggeratedly peaceful and sexy popular portrayal of the bonobo. Firstly we need to recall that bonobo sexual behaviour is not as casually sexy as the authors have stated, and secondly we need to note that chimpanzees have similar sperm competition physiology to bonobos along with aggressive and competitive mating strategies. Both species have male and female hierarchies which affect reproductive fitness: the female's in relation to access to resources and the male's in relation to access to ovulating females.

Male chimpanzee in-group aggression shows that sperm competition *per se* clearly does not remove other forms of male sexual competition. We have also noted that male bonobos can monopolize females in small groups, and that the status of the mother can affect a male's mating and reproductive success in that she supports her son and interferes in the mating of other males. Female bonobos are also able to express more mate choice than chimpanzee females and reject the sexual solicitations of some males which they seem to do especially when most fertile – and sperm competition can only act at that time, when one sperm can actually 'win' by fertilizing an egg and therefore take its 'winning trait' into another generation (for reproductive success not being left to sperm competition see, for example, Gerloff *et al.* 1999, Wroblewski *et al.*2009, Schaller *et al.* 2010, Surbeck, Mundry and Hohmann 2010).

Dominant male chimpanzees and dominant male bonobos (sometimes via dominant mothers) father more offspring than lower-ranking males, even if this is less so than in other species where greater exclusivity of access to ovulating females can be achieved. Matters are more complicated in the chimpanzees and bonobos because of the amount of mating that occurs when females are not actually ovulating but it is only when they are ovulating and the sperm race is actually on that sperm competition can act. Chimpanzee and bonobo reproductive success is certainly not left solely, or even primarily, to sperm competition.

In a study of Taï chimpanzees (Stumpf and Boesch 2004) females were shown to follow a mixed reproductive strategy: selective about mates when conception was likely (proceptive towards some males, resistant towards others) and more promiscuous when conception was unlikely. The authors say that although chimpanzee females are promiscuous, this mixed strategy does not promote sperm competition among males and supports Hrdy's "priority of access" arguments (see below).

Stumpf and Boesch conclude that in multimale groups the reproductive interests of males and females are often in conflict. Females have two possibilities: they may relinquish control over paternity to the males (Ryan and Jethá's solution), or they may actively attempt to influence which males father their offspring. Their study showed that female chimpanzees do not mate indiscriminately but try to influence paternity when most likely to conceive, with a strategy of promiscuity when not ovulating and selectivity during the fertile days of the periovulatory period. (For similar results at Mahale see Matsumoto-Oda 1999)

Keeping in mind this complexity of chimpanzee and bonobo behaviours and how bonobo behaviour is not as relaxed, egalitarian, and sexy as the authors portray, we will now look at the evidence they present regarding sperm competition.

Ryan and Jethá (p. 221) attack Darwin's ideas of female sexual reticence and bring in Sarah Hrdy again, saying that Hrdy doesn't buy Darwin's coy female schtick for a minute, and that she says it "does not apply to the observed behaviour of monkey and ape species at mid-cycle". From the same paper she also states (Hrdy 1996):

"[F]rom a female's perspective, sperm competition is more probably an unfortunate consequence of polyandrous matings than something females were selected to promote. Competition inside her reproductive tract is scarcely the optimal arena for male-male competition, though once such competition gets going in males, females may have no choice but to make the best of it."

"To the extent that genes affect offspring quality, females should fare better under a "priority of access" system (distinguishing between individual males) than under a system emphasizing priority of fertilization (distinguishing between sperm)."

"Sperm competition represents another consequence of male-male competition potentially detrimental to female reproductive interests over the long run... Given that competitive sperm does not necessarily correlate with the most robust phenotypes and that fathers producing competitive sperm need not create advantageous conditions for infant survival, in species like primates where individuals are only long-lived if they survive the very vulnerable infant and juvenile years, I would expect competitive sperm to fall rather far down the "list" of a mother's criteria for an ideal mate. Certainly ejaculate quality does not necessarily correlate with other measures of survivorship or phenotypic success."

Though Hrdy rightly countered the out-dated view that females are invariably 'coy' and monogamous, she is certainly not suggesting benefits to females or offspring from sperm competition.

Ryan and Jethá seem to be promoting a view that female and male sexualities are somehow perfectly matched with little or no difference in motivation for sex. At minimum they present a view that the female can only benefit from simply accommodating the sexual interests of the male. Hrdy, on the contrary, recognizes that what benefits reproductive competition in one sex can be detrimental to the other, and that strategies and counter-strategies evolve in the two sexes in their respective, and often conflicting, reproductive fitness adaptations.

When the apes are compared it is often noted how small the testes of the polygynous gorilla are compared to those of chimpanzees and bonobos. Ryan and Jethá (pp. 222-223) are under the illusion that gorilla testes are "the size of kidney beans" and that they are "tucked up inside the body". It is odd that they should keep repeating this. In their own chart (p. 230) they show gorilla testes as being 29g compared to human testes at 35-50g which would mean that human testes are the size of one and a half kidney beans. Does a kidney bean ever weigh nearly 15g? And though the gorilla testes are not in a pendulous scrotum they are still external and *not* tucked up inside the abdomen.

The position of the testes varies across mammals: the elephant has his in the abdomen, near to the kidneys, whereas other species, such as hedgehogs, moles, and pigs, have other locations in between that of the elephant and that of species with an external, pendulous scrotum. Why there is such variation is still debated and the matter is far from resolved but it is size and sperm production rather than location that correlates with sperm competition (Birkhead 2000).

Human testicular ratios are said by the authors (p. 223) to also be far beyond those of the gibbon, but gibbon testes size is, like that of the orangutan, actually in the same 'ballpark' as human testes size. The authors later (n. 3, p. 341) note that there is a species of gibbon with a pendulous scrotum but they say that "interestingly" it is not strictly monogamous. Unfortunately for the authors, this "not strictly monogamous" behaviour is because this gibbon is sometimes polygynous which should still mean no sperm competition and, according to their argument, no pendulous scrotum (Jiang *et al.* 1999).

To explain why human testes are so much smaller than those of chimpanzees and bonobos (35-50g in humans compared to 118-160g in the much smaller-bodied *Pan*), Ryan and Jethá (p. 226) resort to an argument that human testes must have shrunk *only in the last ten thousand years*. Rather incredibly, to support their argument for this rapid evolutionary shrinkage of testes size, they use as evidence the fact that in species that breed seasonally the testes enlarge in the mating season! In some species of lemurs (which they incorrectly presume to all be small nocturnal primates when many are, in fact, relatively large and diurnal), they state how testicular volume swells up in the breeding season and then shrinks down again. What this variation during the lifetime of an individual has to do with evolutionary change, i.e., the changes in genes and gene frequencies in populations over time, they don't (not surprisingly, as there is no connection) even attempt to explain.

The paper they reference (Pochron and Wright 2002) is actually about a species of sifaka (a relatively large diurnal lemur) where the males with relatively small testes outside the mating season have a greater increase in testes size during the mating season. But, of course, males across species where there is seasonal mating go through all kinds of seasonal bodily changes. Females do too, such as only producing eggs during the mating season which is hardly evidence that a change from non-functioning to functioning ovaries is a sign of how speedily evolution can act on ovaries. The same goes for the increase in body size that males of some species have in mating seasons, or the growth of antlers, or birds producing their ornamental breeding feathers. None of these *seasonal* changes are about how fast evolution acts on any of these traits over generations – the genes are exactly the same throughout these individuals' lives! Seasonal changes are not genetic changes.

Next (p. 227) the authors reference a *Nature* letter (Wyckoff, Wang, and Wu 2000) about the rate of evolution of a gene which acts on the compactness of DNA in the sperm head. Ryan and Jethá do not explain that this is a gene that affects the size and shape of the sperm head and instead they erroneously argue that this is about testicular tissue and that it confirms that testes size responds rapidly to selection pressures.

Next (p. 231) we are presented with men's enjoyment of porn which depicts multiple men ejaculating in or on one woman as 'evidence' showing how men are naturally keen to take part in sperm competition, adding that married men's sexual fantasies are often about being cuckolded. This is too big an issue to deal with here and now, so just a few points will be made. This use of one woman by multiple men removes any individual responsibility or accountability – it is anonymous and without any emotional intimacy which is its attraction to men using opportunistic sexual strategies. Women do not have to be sexually aroused for them to be used in this way, whereas the opposite cannot be the case and therefore one male copulating successively with a number of females cannot be depicted. Men's 'polygynous' fantasies of being the only male 'servicing' multiple females are not fantasies that can be portrayed by internet pornography with its anonymous rapid-fire action functioning to quickly elicit the male viewer's single ejaculation.

Competition, anxiety, fear, and various similar negative emotions can be connected to sexual arousal, probably because they use the same or similar brain pathways because they have been closely interconnected during the evolution of sex down through species. Sexual fantasies are far from being understood, and as female sexual fantasies are often 'rape' fantasies it strongly suggests that we should not leap to conclusions when it comes to these fantasy aspects of our sexuality. Presumably the authors would not argue that women's 'rape' fantasies represent an ancestral sexual preference, though their depictions of incidents of culturally enforced sequential sex experienced by young females, and their arguments for the benefits of culturally enforced 'sharing' of sex, does not exclude that possibility.

~~~~~~~~~~~~~~~~

Ryan and Jethá provide no actual evidence for their view that human testes size has decreased only in the last ten thousand years. The recent mapping of genomes and the comparisons between species does provide evidence that some genes show more evolutionary change than others, and these include genes involved in reproduction and especially genes involved in the immune system.

The way these rates of evolutionary change in genes are measured is by comparing the number of mutations that do not change a particular protein to the number of mutations that do change a protein. Mutations can occur in a gene DNA base pair without changing the actual protein that is made (a 'silent' mutation) or the mutation can change the protein that is made (a 'replacement' mutation) and the new protein will then be open to selection pressures. If the ratio of the mutations that have changed a protein to those that have not changed a protein is relatively high then positive selection for novel proteins is assumed to have occurred. If low, then selection is assumed to be acting against change. Getting down to the nitty-gritty of DNA mutations and selection at this level is still in its infancy so we need to not get too far ahead of ourselves in reaching conclusions.

In the Wyckoff, Wang, and Wu (2000) *Nature* letter (erroneously referred to by Ryan and Jethá as evidence for rapid changes in testes size), they looked at three genes which potentially influence the morphology of the sperm. Compared to the genomes of Old World monkeys, the ratio of 'replacement' to 'silent' mutations was 13/2 for humans, 12/3 for chimpanzees, 8/3 for the gorilla and 31/8 for the orangutan (though they felt less certain of the accuracy of this last result). The highest rate of change was in one of the three genes, with a ratio of 5/0 for both human and chimpanzee and 3/1 for the gorilla (8/2 for the orangutan). Whatever conclusions we might come to about this rate of change and how the apes compare, it is not evidence for a recent rapid reduction in human testes size.

In the same issue of *Nature*, Clark and Civetta (2000) write that the degree of condensation of chromosomes in the sperm head, and therefore its shape, does not seem to be related to the rate of fertilization, implantation, or pregnancy. And Gavrilets (2000) writes about the potential involvement of sexual conflict in the rapid evolution of genes involved in reproduction. He presents a model suggesting that changes in traits are expected whenever females (or eggs) experience loss of fitness such as that caused by *polyspermy* when more than one sperm successfully enters the egg due to there being too many compatible males (or sperms). Polyspermy is responsible for up to 20 per cent of spontaneously aborted human foetuses (Bonnicksen 1991).

If sperm is under selection to out-compete other sperm by placing increasing numbers of sperm in the female's reproductive tract, then the female is under selection to place barriers in the way of sperm to avoid the potential damaging consequence of more than one sperm entering the egg. Whether human females are able to exert any 'cryptic choice' (i.e., choose between sperm internally) in the way that insect females appear to be able to do is unknown (Eberhard 1996).

Though we might expect there to be some mechanism to differentiate between sperm with regard to some information given by their surface molecules, any discrimination between genetic traits is, as Hrdy says, a job better done by the female herself using pre-copulation mate choice. Simply mating with multiple males is only going to select for sperm that is good at sperm competition, and such 'robust' sperm has no necessary link to traits attractive to a female and beneficial to her and her offspring's 'fitness' beyond that 'robust' sperm.

Runaway coevolution, where traits in the two sexes evolve rapidly in response to each other, can potentially lead to reproductive barriers which make only certain males and females reproductively compatible (and thus can potentially even lead to speciation). In sexual conflict the conflict is over the rate of copulation due to the fitness costs females can often experience due to mating, whether due to interference with foraging, to injury, to transmission of pathogens, to immune suppression by ejaculate (and other manipulations of the female via the chemicals in the semen) or due to polyspermy. Not unlike the rapid evolution of the immune system as it is engaged in an 'arms race' with rapidly evolving pathogens, rapid evolution in genes connected to reproduction suggests an 'arms race' between the interests of males (or sperm) and the interests of females (or eggs).

Males and females evolve in response to changes in the other sex so we should also consider how this might be influenced by the *number* of copulations that a female deals with rather than only considering rapid change as a response to copulations with different males. If we take into consideration the fact that in human sexual monogamy females are copulating at a very high rate compared to other monogamous females, how this might affect selection in terms of both sperm production in the male and the defences in the female needs to be taken into account. A high copulation rate even if with just one male would be expected to be responded to by defences in the female's body against what are 'alien' cells, chemicals for his benefit more than hers, pathogens, and the same threat of polyspermy.

Yes, it's complicated. But we'll look at a few more papers on the rapid evolution of genes involved in reproduction to get more of an idea of what is coming to light and to put Ryan and Jethá's argument (and single reference) in context.

First we'll look at a paper by Kingan, Tatar, and Rand (2003) which Ryan and Jethá refer to in a note (n. 5, p. 339). This study concerns one of the semen coagulating proteins, SEMG1. Results showed a 'selective sweep' in chimpanzees which means that there was strong selection for a reduction in the variation in this gene in chimpanzees because of strong positive selection for, and maintenance of, its role in forming a rubbery copulatory plug. Gorillas, on the other hand, showed a loss of function for this gene. Humans showed a nonsignificant reduction in variation though a trend in that direction. So this study shows a difference between humans and chimpanzees, and between humans and gorillas, which fits more with the low levels of sperm competition view of humans, including that of 'the standard narrative', rather than a promiscuous mating system.

Jensen-Seaman and Li (2003) looked at the evolution of SEMG1 and SEMG2. For SEMG1 they note that chimpanzees have an expanded gene which results in a protein twice as long as that in humans. The orangutan ejaculate is somewhere midway between that of humans and chimpanzees in its viscosity. Orangutan females tend to mate when fertile with the one dominant male though probably not exclusively, so their position between humans and chimpanzees suggests low levels of sperm competition in humans.

Dorus et al. (2004) looked at the SEMG2 gene. The rate of evolution measure was 2.52 for chimpanzees and 1.28 for macaques (where females also mate with multiple males). For humans it was 0.91, 0.89 for gibbons, and for the orangutan 0.88. The polygynous colobus monkey came in at 0.70 and the gorilla at 0.61. These figures correlated with the average number of male partners per periovulatory period (1-2 for humans, 8 for chimpanzees) and with residual testis size. So this fits with the current evolutionary biology view of human sexuality, i.e., that human females are not as sexually monogamous as gorilla females but are far less sexually promiscuous than chimpanzees (or bonobos).

Nielson et al. (2005) discuss more generally the genes that are positively selected in humans and chimpanzees. They note that human genes that are targeted by positive selection are significantly more likely to harbour variations that are associated with known genetic diseases. Many of the genes with evidence for positive selection are involved in tumour suppression and *apoptosis* (which is the programmed cell death which can fail and therefore lead to continued cell divisions and tumours), and spermatogenesis. Up to 75% of sperm are eliminated by apoptosis in normal spermatogenesis. The suggestion is that the mutations that arise in sperm cells that enable them to avoid this cell death, and are therefore selected, may have damaging consequences when they are expressed in other cells for which a prolific cell division like that of spermatogenesis is clearly harmful. This could mean that selection pressures to produce more sperm by selecting for the avoidance of apoptosis creates an increased risk of tumours elsewhere. The positive outcome for prolific cell production when it comes to sperm cells can carry with it the negative consequences of avoiding cell death in other body cells.

Clark and Swanson (2005) looked at adaptive evolution in primate seminal proteins and note that seminal fluids show striking effects on reproduction, involving manipulation of female behaviour and physiology, mechanisms of sperm competition, and pathogen defence. They also note a connection between genes showing positive selection and disease, suggesting that adaptive changes in one area can have deleterious effects in another. The prostate is the major source of seminal fluid proteins and they suggest that adaptation in prostate-expressed genes may benefit primate males in their reproductive lifespan but also contribute to disease through their damaging side effects.

Hughes *et al.* (2005) compared the Y chromosome of human and chimpanzee. They found evidence that in the human lineage the unique Y-linked genes have been conserved by selection against negative mutations. In the chimpanzee lineage, by contrast, several of these genes have sustained inactivating mutations. They suggest that this is due to strong selection in chimpanzees on other genes on the Y chromosome connected to the high level of chimpanzee sperm competition. In contrast, our less promiscuous evolution has meant that the human Y chromosome has not suffered the same gene casualties. So the advantages of sperm competition to chimpanzee males has meant that other genes on the Y chromosome, already greatly reduced, have lost out but humans have not experienced the same strong selection for sperm competition so other human Y chromosome genes are still active.

Ramm *et al.* (2007) compared the evidence for positive selection on ejaculate proteins in rodent species to their mating systems, and reanalysed data on the primate genes SEMG1 and SEMG2. For the SEMG1 gene they found no correlation with sperm competition while for SEMG2 there was a correlation with those species with a high rate of sperm competition such as the chimpanzee. Their overall conclusion is that despite the evidence for adaptive evolution in reproductive proteins there is not a simple link between mating systems and evolutionary rates of genes. Many reproductive genes also have immune-related functions, so they suggest that other potential factors, such as host-pathogen coevolution, should not be prematurely discounted as influencing the rate of change of these genes.

Carnahan and Jensen-Seaman (2008) compared a prostate gene, TGM4, in apes and looked for clues to ancestral mating behaviours.

Their conclusion is that the chimpanzee and bonobo ancestors have had a multimale/multifemale mating system going back near to the split with the *Homo* line. Sexual dimorphism and polygyny is also suggested to have a long history in the gorilla, possibly going back near to the split with the lineage leading to chimpanzees and humans. For humans, unfortunately but commonly, clear conclusions are not to be found. The rate of evolution of these genes could mean neutral evolution and the genes are on their way to losing their reproductive competition function. On the other hand they may reflect fluctuating episodes of differing selection due to fluctuating mating systems. This question remains unresolved.

This selection of papers shows that the one paper referenced and misrepresented by Ryan and Jethá is not evidence for a pre-agricultural promiscuous human mating system or for the reduction of human testes size in the last 10,000 years. No one is arguing that humans had an exclusively sexually monogamous past, and the evidence supports rather than contradicts a long human history of low levels of sperm competition. And amidst all the fanfare from Ryan and Jethá for the supposed benefits of promiscuous sex and (p. 227) male "souped-up genitals" and "spermatic firepower", we can again note that evolution acting on male sexual selection does not mean that it is acting for the long-term health of males or for the good of females or offspring.

Ryan and Jethá end their CHAPTER SIXTEEN linking 'gangbang' pornography, the sequential sex providing motivation for the teams of hunters or workers previously discussed, and cheerleaders spreading their legs for football teams. If anything, these male teams – actual or pseudo hunters or warrior 'bands of brothers' – are nothing like bonobos but are *very chimpanzee-like;* male bonobos do not bond.

Imagine for a moment what a *real* bonobo legacy would be: a world without team sports!

The penis

Ryan and Jethá (p. 234) mention copulatory plugs only once and only then in relation to snakes, rodents, some insects, and kangaroos, with no mention of their existence in the chimpanzee and the bonobo. They prefer instead to skip past this and tell us of the interesting design features of the human penis and that humans are "the great ape with the great penis!" They are impressed by the length and the thickness, and particularly by the glans which they tell us is a clever design to remove sperm from other inseminations in the same female. This is based on laboratory evidence using artificial sperm, artificial vaginas, and artificial penises, and as primate sexuality expert Alan Dixson says, there is no credible evidence for the penis as a plunger (Dixson 2009).

Only when a woman is ovulating and therefore fertile can sperm pass from the vagina and through the cervix. The vagina protects itself with immune responses to alien cells, including sperm cells which cannot hang around for long in this harsh (for them) environment before (if the female is ovulating) moving through the cervix. They start to do this within minutes, assisted by changes in the cervical mucus when the woman is fertile; otherwise they leave the way they came in or are broken down by the woman's body. If the penis is going to have any role to play in displacing sperm then copulations will have to follow each other pretty quickly (Suarez and Pacey 2006).

Dixson (2009) also has updates on the relative sizes of ape penises (measurements have actually only ever been few and not particularly reliable). On average the human penis is 15-16cm long when erect, with considerable individual variability. The chimpanzee penis is about 14cm with a range of 10-18cm with similar measurements for the bonobo. The gorilla penis is about 6.5cm – about twice what has been previously thought. As we have already noted, chimpanzees and bonobos have copulatory plugs. The female's sexual swelling greatly lengthens the vagina, and the male's long, thin, 'filiform' penis has evolved to place sperm near to the cervix and to dislodge copulatory plugs left by other males (Dixson and Mundy 1994).

Dixson states that the human penis is not exceptionally long in relation to other primate species, especially when body size is taken into account. In a comparison of 14 species of primates (including a lemur, galago, seven monkey species and the great apes) the human penis in relation to body size does not stand out at all. Only the circumference is unusual, though the spider monkey is at least one other species with a large, thick penis.

The length of the human penis corresponds to the changes in length and positioning of the human vagina and cervix in our evolution. The greater thickness of the human penis is almost certainly linked to the changes due to bipedalism and human childbirth, and the need for a satisfactory fit during coitus (Bowman 2008). Considering the degree to which human females are encouraged to do their 'pelvic floor' exercises, especially after childbirth, not to mention 'corrective' vaginal surgery, the evolved thickness of the human penis seems essential (if not actually on the small side) for human sexual and reproductive requirements.

Sexual selection *has* influenced penile morphology in primates (see overpage).

A-H are examples of primate species with primarily polygynous mating systems.

I-N are examples of primate species with multimale/multifemale mating systems.

Species which have monogamous or polygynous mating systems tend to have a less complex morphology similar to the human penis, including with regard to the glans.

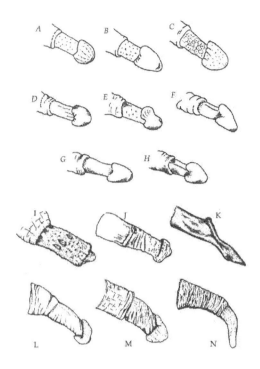

A, B, and C are species of langur monkey.
D, E, F, and G are species of guenons.
H. Patas monkey.
I. Brown lemur; J. Squirrel monkey; K. Stumptail macaque;
L. Crab eating macaque; M. Yellow baboon;
N. Chimpanzee.

Source: From Dixson (2009).
Reprinted with permission of Oxford University Press.

The morphology of the human penis is very similar to (though larger than) that of the gorilla. Comparing the penile morphologies in 48 genera, Dixson concludes that the human penis is not exceptional and is consistent with an evolutionary background of polygyny or monogamy. The rounded or helmet-shaped glans is a phylogenetically ancient trait, present in many Old World monkeys and also in the gorilla (Dixson 2009). The human glans is not a novel trait in humans that evolved for sperm displacement but a common and ancient feature that is also found in species with little or no sperm competition.

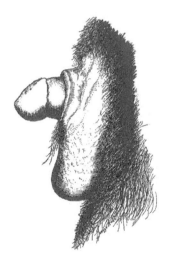

External genitalia of a sub-adult male gorilla
(Gorilla g. gorilla)

Source: From Dixson (2009) after Hill and Matthews (1949).
Reprinted with permission of Oxford University Press
and John Wiley & Sons Ltd.

Sperm production

The rate of human sperm production is lower than that of any other mammal so far investigated. Numbers stored in the epididymis are also low as is the production per gram of testicular tissue. Chimpanzees can copulate once an hour for five hours and still have their sperm store half full whereas men will deplete their sperm store completely with six ejaculations in twenty-four hours (Marson *et al.*1989, Birkhead 2000).

The human sperm takes 74-76 days to produce which is longer than almost any other mammal for which measurements have been made. The sperm reserve of humans is about one tenth that of the chimpanzee, and human sperm counts quickly decrease as a result of repeated ejaculation (Dixson 2009). In one experiment men depleted their sperm stores by ejaculating on average 2.4 times a day for ten days. Their sperm output remained below the pre-depletion levels *for more than five months* (Freund 1963, Barash and Lipton 2001).

Ryan and Jethá (p. 237) view human testes size and sperm production as showing that humans have surplus capacities compared to what any monogamous or polygynous primate would need, again forgetting that no one is arguing for sexual exclusivity, whether by both sexes in monogamy or by females in polygyny. And for human pair-bonded monogamy we also need to consider the effect of a high copulatory rate within the pair bond.

As well as the lack of evidence for changes only in the last ten thousand years, our specific human evolutionary selection over millions of years needs to be factored in, including polygynous mating systems and the high rate of sexual activity within human monogamous pair bonds. The authors include possible evidence for reduced sperm counts in the last few decades which is irrelevant to an argument about sperm counts thousands, and hundreds of thousands, of years ago. Apart from anything else, such recent reductions in sperm counts may even be showing an increase in ejaculatory frequency depleting sperm stores.[21]

Ryan and Jethá (p. 238) write that daily ejaculation reduces the damage to sperm cells. What the study they refer to actually shows is that for men who had a high percentage of damaged sperm, four-fifths had reduced this damage after ejaculating every day for one week. Dr Greenberg, who led the study, also warns that daily sex for too long – say a fortnight – would probably cut sperm numbers too much. It is advised that this daily ejaculation should not be used by men who already have low sperm counts, and the best advice to all couples trying to conceive naturally is to have sex every couple of days.[22]

This is followed (pp. 239-240) by Ryan and Jethá's argument that enforced sexual monogamy for women means that men with low fertility could still pass this to offspring rather than it being selected out as they argue is the case in the chimpanzee. If, due to sexual monogamy, men with lowered fertility were able to sometimes produce offspring they argue that these lower fertility genes would spread, and this, they say, is why human testicles have shrunk.

What Ryan and Jethá seem to be doing here is confusing reproductive *problems* of lowered fertility with testes size and sperm production *per se*. Monogamous and polygynous species with relatively small testes and relatively low sperm production have not been noted to have greater fertility problems; a male would presumably only be at a disadvantage if he were to actually come up against sperm competition involving another male producing more sperm. There has to be sperm competition going on for there to be selection for greater sperm production: no sperm competition, no selection for traits that make the sperm good at competition. As human males all have relatively (compared to chimpanzees and bonobos) small testes and low sperm production there would only be a potential competitive

---

[21] Are sperm counts falling? For some of the controversy over this subject, see:
'What's Going On with Those Scandinavian Sperm?' *Scientific American* guest blog by Emily Willingham, July 11, 2011.

[22] Daily sex 'best for good sperm', *BBC News* online, June 30, 2009.

fertility problem for any one of them if they were actually to engage in sperm competition against more prolific ejaculates.

Actual fertility problems are a reason why hunter-gather marriages might end and new ones be embarked upon. Incompatibility can be a reason why some couples are unable to reproduce together and, as we have noted above, it is sperm competition that can lead to incompatibility between certain males and certain females due to the rapid evolutionary changes that select for increased sperm production in males and improved barriers against that sperm in females.

Human population numbers are hardly suffering, which brings us back to Ryan and Jethá's earlier argument regarding our ancestors' reproductive behaviour. On the male fertility and virility side they are singing the praises of "spermatic firepower" and the battles amongst billions of sperm to be that one winner but on the female and offspring side they are arguing against allowing the survival of any resulting life beyond the replacement level of an average of two children per woman. One would have thought that such high male fertilizing power would not be a particularly useful thing to have in ancestors who deliberately acted against population growth and disposed of all the surplus babies.

In fact, Ryan and Jethá argue positively that our ancestors did have a low reproductive rate because the women had such low fertility which, unlike male fertility, is argued as being a good thing. They argue for very high sperm production along with very low egg production; high male fertilizing power along with few reproductive consequences that spoil the fun. They argue for incredible spermatic and genitalic male-male combat on what turns out to be a near-barren battleground.

It seems a little overly macho to say the least.

Ryan and Jethá (p. 241) note the variation in testes size in different human populations, referencing a *Nature* article by Jared Diamond (1986). This, they suggest, is evidence supporting variation in relation to degrees of sexual promiscuity. They say (n. 13, p. 342) that evidence from the Amazonian tribes that might show signs of sperm competition is not available. But if our pre-agriculture testes were the size of those of our ape cousins it is surprising that such an obvious difference has never been remarked upon by people making first contact with hunter-gatherers across the globe who surely, unaffected by agriculture, should still be sporting testes like those of bonobos and, supposedly, our similarly endowed hunter-gatherer ancestors.

What Jared Diamond (1986) particularly considers in his *Nature* article is the relationship between testes size and the frequency of dizygotic twins (from two separate eggs), the frequency being much lower in Asians than in Caucasians and black Africans. The suggestion is that twins have been selected against in Asians due to smaller body size, and that smaller testis size is therefore a by-product of selection against twins in women. Whether this turns out to be correct or not it is another reminder that selection acts on females too rather than females merely tagging along or accommodating whatever selection produces in the male.

Finally, Dixson (2009) provides further evidence for low levels of sperm competition in humans. The midpiece of the sperm contains the mitochondria which are the energy source for the sperm, and larger midpieces occur in species where females mate with multiple males and testes are larger, i.e., where sperm competition is occurring. The midpiece in human sperm is relatively small compared to other primates – including the orangutan and the gorilla. Also, the vas deferens, which has peristaltic muscle contractions to transport sperm during sex, has the longer length and lower thickness in humans that is typical of species with low sperm competition. And finally, the oviduct in females is longer and more convoluted in species that have more sperm competition; the human oviduct is relatively short, again pointing to a low level of sperm competition.

There are, therefore, many physiological differences beyond testes size that are evidence for a low level of sperm competition in humans, and this all points to the conclusion that changes in our mating system happened long before the advent of agriculture.

Prostate cancer

Ryan and Jethá (p. 238) note that trace amounts of carcinogens are often present in semen so there is a suggestion that a regular 'flushing out' may help prevent cancer in the prostate, though I'm not sure how much this is meant to encourage women to be the recipients of these carcinogens. Whether some particular rate of ejaculation helps prevent prostate cancer is still controversial and it is very unlikely to be a significant preventative measure. There has, for example, also been some evidence that a virus can be a cause of this cancer, especially a cause of the more aggressive tumours (Schlaberg *et al.* 2009). If a sexually transmitted virus is a potential cause then sex with multiple partners is not going to help.

Ryan and Jethá certainly like to give the impression that if what they argue are naturally evolved behaviours are given free rein then no harm can possibly come of it; it is as if they believe that evolution acts primarily to take care of individual organisms. With a high sex drive, especially in males, the implication is that this can only be a force for 'good', including long-term health and well-being. Earlier we noted aspects of sex in nature which are certainly not what we would associate with 'goodness' or well-being, whether rape in mallard ducks or the sorry short life of the antechinus male. So what about the prostate?

Secretions from the prostate constitute about thirty per cent of human semen. Summers and Crespi (2003) give a general review of the connections between sexual selection, androgens, and the prostate gland before focusing on the androgen receptor. Androgens, such as testosterone, are known to influence aspects of male reproduction and are likely to be involved in sexual selection. As a consequence of sexual selection, traits that increase male reproductive fitness relatively early in life can have negative effects on survival later in life. Natural selection is about increasing copies of genes in the next generation, and if this is best achieved by a short period of high reproductive success and then a relatively early death then traits that promote reproductive success during that short period but lead to a shorter lifespan are selected.

This is why males across species generally have a shorter lifespan than the females because a 'live fast die young' strategy in males can produce high numbers of offspring within a short period whereas females maximise their offspring numbers by having a longer reproductive timespan. Monogamy tends to equalize the lifespans of the sexes because it removes the benefits to males that come from monopolizing a number of females over even a brief period. If males are also important as parents then they too come under the same selection pressures as females to be in it for the long-haul.

In humans, increased androgen levels are associated with the risk of prostate cancer, amongst other costs (Bribiescas 2001, Soronen *et al.* 2004). Variation in genes involved in androgen metabolism and cell proliferation in the prostate show an association with the risk of prostate cancer. Virtually all cases of prostate cancer involve mutations in the androgen receptor, and variations in the gene for the androgen receptor are associated with cancer risk.

Summers and Crespi looked at a section of the gene which when short is associated with cancer and when it is long is associated with infertility, low rates of spermatogenesis, and reduced expression of male-specific traits. These findings suggest a trade-off between benefits in reproduction and long-term health: higher fertility with higher rates of sperm production appears to also carry greater risks of cancer.

The androgen receptor is also expressed in a variety of tissues in females and appears to be linked to some kinds of cancer such as ovarian cancer where androgens increase cell proliferation and decrease cell death indicating a cancer-promoting effect. What is good for cells such as sperm that might benefit from a fast rate of cell division and an avoidance of any programmed halt to this proliferation of cells can have consequences elsewhere where such a proliferation becomes a harmful tumour. There can be strong positive selection in one area which carries with it the risk of tumours elsewhere.

It is a poor understanding of evolution and natural selection that leads to assumptions that sex, and all things connected to sex, bring health and well-being to the individual. Bodies are all ultimately sacrificed for the benefit of the genes they carry, and selection favours the spreading of those genes over the health and well-being of the body that is their temporary home. Bodies are selected to serve the needs of the eggs and the sperm above all else. It may feel as though evolution and the 'naturalness' of lots of sex ought to promote health and 'fitness' of the body but the 'fitness' is that of the genes spreading through a population down through generations. We can see, though, how genes can benefit from creating minds, thoughts, and behaviours in their temporary home that lead to their spread even as they simultaneously harm the long-term health and well-being of their temporary home.

Natural selection acts by 'selecting' certain genes over their alternatives. 'Fitness' concerns the relative number of particular variants of genes which succeed in spreading down through generations. If individuals become diseased or cancer-ridden when they are past their prime reproductive period then, if they have spread their own gene variants, they are still evolutionarily 'fit'.

There is also increasing evidence that males face a trade-off between sperm quality and immune function. Sperm cells are 'non-self' (alien cells) to the male and are therefore targets of immunological attacks in the male's own reproductive tract. Physiological mechanisms partly shelter sperm production from immunological attack but not completely, and adverse effects on sperm quality can occur. High quality ejaculates require a wider suppression of immune responses which can be costly to the individual, especially those with lower genetic resistance towards infections (Skau and Folstad 2004).

Lewis, Price, and Weddell (2008) also review aspects of sperm competition including how it affects male immunity and cancer. They note that several mammalian genes are expressed at very high levels only in rapidly dividing embryonic cells, spermatogenic cells, and malignant cancer cells, and that there are at least 88 genes expressed only in testes and malignant cancer cells. 'Apert syndrome' in humans, which is an abnormality of the skull, face, and limbs, is usually caused by a mutation which also increases the rate of replication of cells that produce sperm cells. Lewis *et al.* conclude that it is now clear that there can be a direct link between genes regulating immunity and male ejaculate quality, and there appears to be a link between the demand for the production of many sperm and the risk for cancer, and possibly other diseases.

Marlene Zuk in *Riddled with Life* (2007) writes about the health risks of being male. As well as the sex differences in behaviour across species which can affect survival, such as distances males might travel or physical aggression, Zuk also discusses the connection between testosterone and a weak immune system, noting that a wide variety of vertebrate species have males that are more susceptible to disease. Indirectly testosterone can cause the levels of other hormones, such as cortisol (one of the so-called stress hormones) to increase which depresses the immune system further. In women, on the other hand, it is the autoimmune disorders that are more severe – the result of an over-vigilant immune system.

In evolution, short-term gains have the advantage over long-term investments, so dying young might be expected under certain selective conditions. For males, producing so many sperm, it is possible to concentrate reproductive success within a relatively short time-frame when they are at their prime and able to out-compete other males. For females, their reproductive success depends more on the long-haul, producing their limited number of offspring over the longest time possible.

Zuk discusses this sex difference, also using the antechinus as an illustrative example. She writes that no female antechinus will suffer from the lack of a male, and the female may produce at the very most 36 offspring from three litters over three years, though most females never do so. A male that can out-compete other males can potentially produce litters with more than three females in one mating season but this is balanced with the other potential male outcome of producing no offspring at all. For males, selection acts differently than it does for females; if males die relatively young due to the risks and behaviours that also lead to more offspring for some, then mating behaviour wins out over long-term survival.

~~~~~~~~~~~~~~~

The breeding system of our ancestors since our divergence from our common ancestor with the chimpanzee has no doubt changed and probably fluctuated through time. The direction selection takes depends on what it has to work on to start with (such as the particular gene pool of a founding population), what mutations arise, and other random factors, but the traits that enable members of the same sex to out-reproduce each other will be a major selective factor.

Ryan and Jethá are not only wrong about the level of sperm competition in our pre-agricultural ancestors, they are also wrong about sex and selection acting for the benefit, well-being, long-term health, and overall lifespan of the body that produces the gametes, especially when those gametes are sperm.

*

- Sexual size dimorphism needs to take into account selection pressures acting on both sexes, not only males.

- The fat-free comparison of human males and females points to a polygynous breeding system; on the other hand, our sexual division of labour can account for our sexual dimorphism within a monogamous breeding system.

- Ryan and Jethá's argument that polygyny depends on the accumulation of wealth and political power that only came with agriculture totally disregards the widespread existence of polygyny across hunter-gatherers and the benefits men gain from these marriages.

- Sperm competition has not replaced other forms of sexual competition in chimpanzees or bonobos.

- Seasonal change in testes size is not evidence for rapid evolution of testes size any more than seasonal change in other traits, such as antlers or ornamental feathers, shows that they experience rapid evolution.

- Ryan and Jethá provide no actual evidence for their view that human testes size has decreased only in the last ten thousand years.

- Rapid evolution in genes connected to reproduction suggests an 'arms race' between the interests of males (or sperm) and the interests of females (or eggs).

- The length and the thickness of the human penis correspond to the evolutionary changes in the human female and are not indications of promiscuous sexual behaviour.

- If a reduction in human testes size is due to agriculture then it needs to be explained why humans unaffected by agriculture have also seen the same reduction.

- There are other physiological differences – sperm midpiece, vas deferens, oviduct – that are evidence for a low level of sperm competition in humans, all pointing to the conclusion that changes in our mating system happened long before the advent of agriculture.

- Traits that are selected because they increase reproductive success can also have negative consequences regarding the long-term health and well-being of the individual. This is usually more so in the case of males and there is evidence that traits such as those selected that favour sperm production can carry negative consequences for the immune system and can lead to tumour growths.

Male sexual motivation is relatively straightforward: from the first males that expelled millions of sperm in the vicinity of eggs in the early oceans, expelling millions of sperm in the vicinity of eggs has been a strongly selected behaviour. It comes as no surprise that men tend to be rather obsessed with their sperm production and their means for expelling those sperm in the vicinity of eggs – serving sperm has been the main shaper of male bodies and behaviours.

Females have evolved from expelling millions of eggs in the vicinity of sperm to something quite different in many species. Reduced numbers of eggs, internal fertilization, gestation, lactation, parenting, and so on, serving eggs has come a long way from a simple expulsion of them in vast numbers.

So, what about females?

CHAPTER NINE

Let's Hear it for the Girls
(Though mostly it's still about the boys)

Ryan and Jethá (p. 245) contrast the male's speedy road to ejaculation with the longer time it takes for women to "get warmed up" and then: "after an orgasm a woman may be anticipating a dozen more".

We are familiar with the exaggeration and lack of honesty from men with regard to, for example, the size of their penis, their number of sexual conquests, or their sexual performance, so perhaps we should be prepared for something similar from women too. In the battle of the sexes when it came to women rightfully arguing against their supposedly passive and monogamous sexual nature, exaggeration of the female's sexual capacity probably seemed like an effective strategy to counter the sexual bravado of the male. The potential for multiple orgasms is something men do not have (though some may achieve them through non-ejaculatory practices) so we might expect women to exaggerate this ability when the reality may be that it is not usually a realized potential and few women expect, enjoy, or even want more than one good orgasm, never mind a dozen.

Ryan and Jethá think that the mismatch between the sexual responses of men and women is simply explained by their argument that human females have evolved to mate sequentially with multiple males. How, they ask, can men and women who have evolved together as sexually monogamous couples "for millions of years" end up so incompatible? How can we have ended up with such "flagrantly maladjusted sexual responses"? They are under the mistaken impression (because they have ignored other species) that evolution cares about matching the interests and behaviours of the two sexes.

By now it should be clear that the evidence points decidedly to humans having evolved a multifamily social and breeding system probably within the last million years or so and certainly well before agriculture. The numerous aspects of our own evolution already discussed should by now be enough to show that the simplistic *Sex at Dawn* story is devoid of genuine evidence, and why, when it comes to sex, we (Ryan and Jethá p. 246) "turn away from chimpanzee and bonobo models", not, as they say, "prudishly" but *evidentially*.

Ryan and Jethá (pp. 247-249) discuss the history of the treatment of the female "disease" of *hysteria* when it was really nothing more than female sexual frustration and (p. 248) they say that such treatment was "one element in an ancient crusade to pathologize the demands of the female libido – a libido that experts have long insisted hardly exists". They then discuss how eliciting an orgasm as a "medical" treatment for hysteria ran alongside the maintenance of the conviction that "female sexuality was a weak and reluctant thing", and that self-masturbation was not acceptable, sometimes even "cured" by clitoridectomies (p. 251).

These attitudes towards female sexuality have existed and are undeniably wrong by our standards but they have not been the only attitudes. For example, in Tudor and Stuart England female orgasm was thought to be necessary for conception, and orgasm was thought to provide an incentive for women to risk pregnancy. Many works discussed the clitoris as the locus of pleasure, and even that women with strong sexual desires were less likely to suffer disorders of the menses. So, as well as the negative attitudes towards female sexuality and orgasm there have been others recognizing their normal existence and healthy nature (Maines 1999).

But more relevant to our discussion here of the link between the negative attitudes to female sexuality and the rise of agriculture in the last 10,000 years, we need to recall the attitudes to female sexuality in a number of the partible paternity societies we looked at earlier. Whereas Ryan and Jethá were keen to tell the reader of the multiple extra-marital sexual relationships and sequential sex, they forgot to mention the attitudes to the female's sexual responses. The Canela, for example, do not allow masturbation for either sex, females are meant not to move during their *very* brief sexual encounters, hands are not allowed to touch genitals, and female orgasm is not often, if ever, experienced. For comparison it would be interesting to discover how these numerous societies – societies that are discussed by Ryan and Jethá to support casual attitudes to sex and paternity but where female sexual enjoyment is mostly a non-issue – deal with their female "hysteria".

In actual hunter-gatherers, such as the Aka, female orgasm is expected and experienced at least some of the time. As we have also noted when looking at hunter-gatherers, sexual fidelity within monogamous or polygynous family units is very important, so sexual monogamy and sexual pleasure are certainly not mutually exclusive as Ryan and Jethá try to argue; on the contrary, mutual sexual pleasure is a very important aspect of these monogamous marriages.

Female copulatory vocalization

Ryan and Jethá's CHAPTER NINETEEN begins with a discussion of female copulatory vocalization. They tell us that women make loud noises during sex "from the Lower East Side to the upper reaches of the Amazon", overlooking the fact that as far as the Amazonian tribes are concerned, signs of female sexual enjoyment are sometimes discouraged and the existence of the female orgasm is often not even recognized. In all of these tribes where we have had the information, female sexual pleasure is either a non-issue, discouraged, or sex is as private and as quiet as possible.

We are told by Ryan and Jethá (p. 256) that the sounds women make during sex are because, like some other primates, our pre-agricultural foremothers wished to incite all the men within hearing to come and have sex with them too. To the authors it just does not make sense for a woman to call attention to herself when mating if she does not want this to happen. It is strange, then, that those tribes where there *is* sequential sex the girls are not even allowed to move, never mind make a noise, and that all other sexual encounters in these and hunter-gatherer societies are so private. Most sex is relatively quiet if others are around, or the couple go into the forest, woods, or bush for privacy – especially if these are extra-marital liaisons. There is no evidence from these societies to support a view that noisy sex by modern Western human females is a legacy of pre-agricultural ancestors and a mechanism to attract multiple other males for sex.

Maestripieri and Roney (2005) looked at studies of female copulation calls in primates and they say there are at least 15 different hypotheses that have been proposed to explain them. One hypothesis is that copulation calls are used by females to encourage mate guarding by their preferred mating partners and reduce the likelihood of sperm competition. There is a lot of variation in the use of these calls across even closely related primate species and it is overly simplistic (to say the least) to assert that they show our female ancestors before ten thousand years ago were using these calls to attract all the other males and that our ancestors were involved in open sperm competition.

One factor about sperm competition which the authors have overlooked is that it only acts when the female is actually fertile and an egg is actually fertilized. If females are seeking multiple males for sperm competition then it does not explain why they would scream for them to come along when they are not actually ovulating and sperm competition is completely out of the question. If there is selection for copulating outside of oestrus, whether with one or multiple males, it is not because of sperm competition. As the authors have argued, ancestral women would rarely have been fertile – rarely ovulating – either because they were already pregnant, lactating, or underweight, so ovulation, and therefore actual potential sperm competition events, would be quite rare in our female ancestors.

Though female chimpanzees do mate with many males when they have their sexual swellings they sometimes have no choice due to male dominance over females and their use of aggression and intimidation to ensure female sexual compliance (Muller, Kahlenberg, and Wrangham 2009). And as we noted in the previous chapter, female chimpanzees can show a mixed strategy of promiscuous mating when conception is unlikely and much more selective mating when conception is likely (Stumpf and Boesch 2005). Bonobo females seem able to express even greater choice especially when ovulating.

Copulation calls in chimpanzees and bonobos are proving to be far more complex than merely serving to elicit sperm competition. A recent study (Townsend, Deschner, and Zuberbühler 2008) of wild female chimpanzees found no support for the sperm competition hypothesis: females did not produce calls when mating with low-ranked males so it was not about inviting other males to the party, and calls did not correlate with fertility and the likelihood of conception. Chimpanzees produce copulation calls at a much lower rate than other primates suggesting they use them as a social strategy. In this study the females called significantly more while mating with high-ranked males, but suppressed their calls if high-ranked females were nearby. This suggests that females attempt to avoid being the target of aggression from other females and also attempt to secure future benefits from the socially important males.

The authors of this study say that there is increasing evidence from a number of chimpanzee communities studied in the wild that female competition plays an important role in dictating female behaviour. Copulation calls appear to be a strategy to advertise receptivity to, and matings with, high-ranking males, confuse paternity, and secure future support from these socially important individuals. Rather than it being a simple case of female sexual desire for multiple males and the inferred reproductive benefits of sperm competition, this study suggests that copulation calls are more about gaining support from high-ranking males and avoiding competition from other females which is a significant risk factor for wild female chimpanzees.

Bonobo females do not experience the same degree of aggression from males, they do not experience such an intense periovulatory period, and they can be more choosy especially when most likely to conceive. Maestripieri and Roney (2005) say that it is not clear whether the "nasalized screams" occasionally given by bonobo females in the context of mating are acoustically and contextually similar to the copulation calls of other species. Clay and Zuberbüler (2012) write that the single or succession of high-frequency squeaks and screams that sometimes occur during copulation are distinct vocalizations not observed in other bonobo interactions, so they are only copulation calls. They found that females were more likely to call with male rather than female partners but the patterns of call usage were very similar in that females called more with high-ranked partners (as in chimpanzees), regardless of the partner's sex. With a female partner copulation calls were consistently produced only by the lower ranking of the two females, showing that call production cannot be explained by physical stimulation alone.

In agreement with what was discussed in Chapter Three, this study also shows that low-ranking females engage most frequently in sexual interactions, both with low- and high-ranking partners, while genital contacts between two high-ranking females are very rare. The presence of the alpha female in the audience increased the likelihood of copulation calls which is in contrast to chimpanzees; while chimpanzee females inhibit call production to reduce risks of sexual competition from dominant females, in bonobos the increase in calling in the alpha female's presence suggests that the low-ranking female is advertising the fact of having been selected for sex by a socially more established female.

Although low-ranked females generally remained silent during interactions with other low-ranked females, they responded strongly to solicitation by a high-ranked female. High-ranked females during the same interaction remained silent. Genital contacts between bonobo females are used as a means to express social dominance relationships because, unlike chimpanzees, bonobo females do not appear to have a formal vocal signal of submission, greeting, and willingness to interact. Clay and Zuberbüler write that interactions with high-ranking females can be dangerous as aggression typically occurs down the female hierarchy so copulation calls appear to help females to advertise their sexual interactions, especially with high-ranking partners. The copulation call is a behaviour that is part of a broader strategy to form associations with socially important group members.

Ryan and Jethá take a very simple hypothesis about copulation calls serving to elicit sperm competition and apply it to our pre-agricultural ancestors and indirectly to ourselves today. But even when we look at chimpanzees and bonobos we find that this simple answer does not explain their copulation calls either, and that rather than encouraging sperm competition they function to signal social dominance relationships. Even if copulation calls did originate as a means to elicit sperm competition in these species they have evolved to have these important functions in social dominance relationships so it is untenable to argue that they merely have some primitive function to elicit sperm competition in humans.

Further, these studies of chimpanzees and bonobos bring up increasing recognition of the significance of female-female competition and female social and sexual strategizing – things that are glaringly absent from the andro-phallo-ejaculo-centric story of *Sex at Dawn*.

We have seen that sex in humans is most often a very private activity. We have also seen (both in the South American tribes and African hunter-gatherers) that female sexual pleasure most often occurs within a long-term, usually marital relationship suggesting the great importance of mutual sexual pleasure for bonding in marriage.

Verbal communication during sex is not usually the easiest of matters, and women do respond to sex differently from men, so men, if they care, can benefit from knowing how aroused the female is and when she does reach a climax; vocal signals during sex enable the human female to communicate what she is experiencing to the male. The man's efforts can show how much he cares about her, which is likely to indicate to her whether he has more than just an opportunistic sexual interest in the relationship, while her vocalizations signal her satisfaction with him and that she is not likely to be seeking other men.

Human female copulatory vocalizations, if they occur, are normally either only heard by the woman's partner or they are emitted in a situation where she is safe from the attentions of other males (or, for that matter, other females) who may hear. Rarely is she ovulating so sperm competition, as in chimpanzees and bonobos, is not what it is about in humans. If this noise is a legacy from our past it has certainly evolved in us at least as much as in our ape cousins and, taking all the evidence into consideration, functions in our own particular pair-bonding communication in our own particular human relationships. Modern pornography, the acting of sex workers, and the very low threshold for male sexual arousal does not alter the fact that sex in hunter-gatherer societies presents evidence only for low-level sperm competition.

The female breast

Ryan and Jethá (p. 259) come now to the human female breast and its "amazing power". We have already mentioned this in Chapter Seven in the context of their argument that our female ancestors were calorie-restricted; at this point in their book the authors now argue that rather than calorie-restricted our female ancestors used their curves to wield power over men. The authors present their evidence of the "Venus of Willendorf", one of the oldest human images and created about 25,000 years ago, which is certainly not evidence for calorie-restricted females:

Venus of Willendorf
(Source: Matthias Kabel)

Ryan and Jethá dismiss utilitarian interpretations for the human female breast saying that the fatty tissue has nothing to do with milk production so there must be other reasons for these costly appendages. They (p. 260) skim over arguments that breasts may advertise fertility or store fat deposits necessary for pregnancy and lactation so that they can give time to the theory they apparently prefer: "the genital echo theory". This is a theory that the breasts evolved to echo the buttocks as a sexual signal after we became bipedal, i.e., breasts exist "to provoke the excitation males formerly felt when gazing at the fatty deposits on the buttocks". They then mention how interest has repeatedly moved between breasts and buttocks, the similarity of the breast and the butt cleavages, and that breasts most probably are fertility signals as both breasts and fertility fade with age.

With an assumption that our female ancestors had sexual swellings like those of chimpanzees and bonobos, the genital echo theory argues that when we became bipedal this sexual stimulus moved to the chest of our ancestral females. Firstly, we cannot simply assume that our last common ancestor with the chimpanzee and bonobo had sexual swellings. Sarah Hrdy (2000) says that while many people still assume that our hominin ancestors had sexual swellings this is unlikely and sexual swellings most likely evolved in the line leading to chimpanzees and bonobos. Secondly, the sexual swellings of chimpanzees and bonobos are fluid-filled and not fatty deposits; so without these fluid-filled swellings there are neither buttocks nor fatty deposits. The muscles that create the buttocks evolved as a consequence of bipedalism, and fatty deposits on the buttocks of the female evolved some time later when the demands of human infants selected for these reproductive fat stores in females along with those of the breasts.

So there were no swollen, fatty female rears that would have provoked male sexual excitement; bipedalism came first, then the muscles, then the fat. The fat stores of the breasts may also have only come later in our evolution, though there is some reason to believe they came earlier. In other primates the breasts are only swollen when the female is nursing an infant which, rather than a sign of fertility, is a sign that the female cannot conceive. If our female ancestors were selected to conceal ovulation and to appear to be constantly fertile then permanently swollen breasts may have acted to remove swollen breasts as a signal of infertility.

Another factor about the human female breast we might consider concerns the needs of a nursing infant. Earlier we noted the problems an upright mother ape would have in transporting an infant that could no longer travel on her back as it could when she moved quadrupedally. In other primates an infant can grip the mother's fur and position itself to feed at a nipple that is close to the mother's chest wall, so our female ancestors as bipeds that also lost their body fur had infants unable to grip onto their mothers and access the nipple. Nipples that moved away from their position against the chest wall would be more easily accessible either to an infant held in the crook of the mother's arm or lying with her as she rested during the day or slept through the night. The breasts of elephant females are remarkably similar to human breasts and no one would argue that they evolved for the pleasure of the male rather than for the nipple to reach the mouth of the nursing calf.

When considering adaptations that relate to females and reproduction it is important to at least consider those that are advantageous to the offspring rather than limiting our thinking only to the sexual interests of the adult males.

What we do not get in the other primates are the female curves due to reproductive fat which is a unique human trait. It is also probably a relatively recent trait within the last million years or so, from when our infants became such resource hungry dependents. As bipedal primates the human female body shape, with the relatively narrow waist along with fatty breasts, hips, and thighs, was selected to enable adequate locomotion in an upright body that would experience the changes due to pregnancy and highly dependent nursing offspring (Pawlowski and Grabarczyk 2003). Human males are attracted to this particular human female body shape because it is a shape that signalled female reproductive success.

Ryan and Jethá (p. 261) mention the gelada baboon, a "vertically oriented primate with sexual swellings on the females' chests". Geladas mostly eat grass in a sitting position and shuffle along so this is believed to be the reason for the sexual signals of the females having relocated to their chest. Both sexes have the pink or red coloured hourglass shape on the chest: in males it can signal testosterone levels and status, and in females the periphery of it 'blisters' with fluid-filled beads of skin when she is fertile. Interestingly, though, geladas are one of the species mentioned earlier that live in very large multimale-multifemale herds made up of polygynous breeding units, and males are about fifty per cent bigger than females. Having a sexually attractive chest area does not, therefore, signify promiscuous mating.

And finally, as far as the similarity between the cleavage of breasts and buttocks goes, the cleavage of the breasts is only achieved with push-up bras which, to the best of my knowledge, our female ancestors did not wear.

We all like curves. We like the curves of our chubby human babies, including their cheeks, their bottoms, and their creased, chubby thighs. We like the curves of men's buttocks and those of their upper arms – all the curves that male muscles make. The breasts of human females signal that they old enough to be fertile, and the size and shape can give other information about age, health, hormones, and fat stores – important information about potential reproductive success. None of this indicates, as Ryan and Jethá (p. 259) say it does, "millennia of promiscuity and sperm competition".

The female orgasm

Not surprisingly, Ryan and Jethá (pp. 262-263) dismiss a role for the female orgasm in monogamous relationships. They prefer the idea that it is to encourage females to mate with every male they can and then (p. 264) to aid in sperm-sorting within the body, assisting the best sperm to get to the egg. These internal assessments, they say, may go well beyond general health traits to the subtleties of immunological compatibility. This immunological compatibility is indeed possible as there are important immunological benefits to offspring in getting a good varied mix of immune-function genes and especially to avoid having eggs fertilized by close relatives; even plants have mechanisms for this type of sorting of pollen (Blum 1998).

But plants are sessile – they cannot move to avoid pollen, even their own, and self-pollination would remove the point of sexual reproduction. For humans, as in mice, smell can act to provide information about what are called the major histocompatibility complex genes (MHC) and numerous studies have shown that women find the smell of different men either attractive or repulsive in relation to their MHC gene compatibility. This, though, suggests mechanisms of *pre-copulatory* mate choice rather than leaving it to the female reproductive tract. Ryan and Jethá do later (p. 275) mention these studies but do not give any acknowledgement that pre-copulatory mate choice matters in women.

Ryan and Jethá (pp. 264-265) do not support pre-copulatory female mate choice and tell us that "a man who *appears* to be of superior mate value...may in fact be a poor genetic match for a particular woman" (emphasis in original). So, they say, the woman may benefit from sampling many males and let her body decide as her body "might be better informed than her conscious mind". In Chapter Six we noted their argument regarding the 'disconnect' between what women think or say they feel about their experience of sexual jealousy and what their bodies' physiological responses are. Now we get to their "crucially important point" about women's bodies being better than their minds in choosing between sperm.

Their argument is one for the equalization of male access to women and the removal of conscious female mate choices, therefore ending the sexual rejection experienced by most males. In complete contrast, women at no point are argued as all being equally attractive to men, and the authors' discussion of women's bodies and sexual signals strongly suggests that they do recognize that men have quite strong mate preferences for young, fertile, and attractive women. The *Sex at Dawn* argument is about men of all ages and ranges of attractiveness getting access to the most desirable female bodies, i.e., that the sexes are equal but one sex is more equal than the other.

This sperm competition argument by Ryan and Jethá is also a very 'reproductive' one, as if we only have sex and female orgasms when an egg is waiting and sperm competition can be in action. Human females do tend to be more sexually proceptive, more spontaneously aroused, and can more easily achieve orgasm when ovulating. How often would our female ancestors have been ovulating? Rarely. They would not have been ovulating once pregnant and then not for a few years when lactating and even then not if they were calorie-restricted. Even when experiencing menstrual cycles they would only be 'fertile' for about five days of each month. So how would the authors' explain all the sexual activity and female orgasms with the come-hither screams to facilitate sperm competition when there are no eggs over which to compete?

The authors' other argument is for the supposed social-bonding role of sex but how many ancestral males would choose to 'socially-bond' with a middle-aged or post-menopausal female when there are younger alternatives screaming out for 'social-bonding' elsewhere – young females who, according to Ryan and Jethá, were not letting their minds get in the way of all that 'social bonding'.

Back in the real world, as we have seen, the extensive human social networks are enabled by pair bonds, and our extra-marital sex is mostly about females acquiring meat or other resources and occasionally about men bonding by 'sharing' women for their own interests with little if any concern for female choice or female well-being.

Ryan and Jethá (p. 265) say that the complexity of the human cervix suggests it evolved "to filter sperm of various males". To support this they then quote information from Dixson (1998) on the especially complex cervix of macaque species, distorting the quote in such a way as to imply a close similarity to the human cervix. Dixson initially is only discussing studies of the cervix of some macaque species and how the very complex macaque cervix might function as a reservoir for sperm. Whether, he adds, the macaque cervix provides an arena for cryptic female choice remains a matter for speculation as the macaques are exceptional in this regard; other multimale-multifemale species with sperm competition can have a simple cervix with a straight canal – the female chimpanzee, for example.

Dixson then suggests that cervical mucus may be more important than the shape of the canal and he only then mentions humans and the change in human cervical mucus around the time of ovulation. The mucus may, he says, filter out morphologically abnormal sperm and this is where Dixson ends with the suggestion that the evidence about macaque and human cervical *mucus* indicates a filtering mechanism and a temporary reservoir for sperm. There lies the similarity.

Dixson is *not* saying that the human cervix has the same complexity as that of the macaque, only that the cervical mucus may act to filter sperm. This does not have to be sperm from multiple males; it is only about filtering morphologically normal and abnormal sperm. We have already learned how the female body acts to reduce the number of sperm that reach the egg and the risks of polyspermy; sex with just the one male presents enough of a problem in this respect. If anyone remains under the illusion that primate sexuality expert Alan Dixson supports the argument for ancestral human promiscuous mating they should read his 2009 book: *Sexual Selection and the origins of Human Mating Systems* and his interview for the *New Zealand Listener*.[23]

Ryan and Jethá (p. 266) continue with their argument that if "female choice (conscious or not) can happen *after* or *during* intercourse" (emphasis in original) rather than before, "helping along those [sperm cells] of a man who meets criteria of which she may be totally unaware" then Darwin's "coy female" starts looking like "what she is: an anachronistic male fantasy". Apparently, a female who finds all men equally attractive, rejects none, is happy with a succession of quickly ejaculating men, and does not use her brain to make any choices is not a male fantasy! In reality, aren't the 'Madonna' *and* the 'slut' both male fantasies?

---

[23] 'Sex Wars: A Wellington professor takes issue with the arguments about monogamy in *Sex at Dawn'*, Rebecca Priestley, *New Zealand Listener*, August 21 2010. http://www.listener.co.nz/current-affairs/science/sex-wars/
retrieved April 01, 2012

Ryan and Jethá then make an unreferenced statement that women's orgasms provoke changes in vaginal acidity which can favour sperm that arrive with the female's orgasm. Though female sexual arousal may slightly alter the acidity of the vagina it is the alkalinity of the semen that neutralizes vaginal acidity and protects the sperm regardless of whether a female has an orgasm or is even sexually aroused (Suarez and Pacey 2006). Fox *et al.* (1973), for example, found no change in vaginal acidity if a condom was used.

~~~~~~~~~~~~~~~~~

The only other addition to Ryan and Jethá's cursory treatment of female sexuality is in their brief CHAPTER TWENTY where they tell the reader about the female's "erotic plasticity". This is a term coined by Roy Baumeister whose 2000 paper is the reference given by the authors.

Baumeister writes that 'erotic plasticity' is greater in women than in men, and women are more able to adapt to changing circumstances. He says it also means that women may end up doing things that are not in their best interest and they can be more easily talked into doing something they don't really want to do or is not good for them. Being receptive to external influence can therefore, he says, also mean gullibility. He gives three hypotheses why there is this sex difference: the physical, socio-political, and economic sex differences give the male the advantage in getting his own way so female plasticity is an adaptation to male power; secondly, the female default mode is to reject sex so she has to be convinced to say "yes"; thirdly, the plasticity is due to the female's weaker sex drive, and a relatively weak motivation is presumably more easy to redirect, channel, or transform.

Baumeister says that there is often a mismatch between when a woman wants sex and when she has sex so she has to be flexible enough to respond positively and competently to sex when she does not particularly want it. Distinguishing between female sexual receptivity and desire is important and he notes that female sexual behaviour does not correlate most strongly with female sexual desire.

He also notes that females, due to their reproductive costs and their limited number of offspring, should be selective about partners which would be expected to lead to lower plasticity. But, he says, selectivity mandates a complex, careful decision process which attends to the particular situation, and it is this situational aspect which leads to the greater plasticity in female sexual behaviour.

So here we have something that fits with female sexual behaviour often not being about sex *per se*, as it so straightforwardly is in the male, but depends on the situation and the costs and benefits to the female of sex in that particular situation.

In a later paper Baumeister (2004) says that of his three hypotheses he considers the female's weaker sex drive to be the most likely explanation. He says that it is weaker motivations in general that lead to greater plasticity of responses. For example, parenting drives and motivations are weaker in men so there is a greater plasticity of the father role. In contrast, mothers have strong motivations for parenting so there is less scope for "compromise, transformation, or modification" of maternal behaviours. With regard to sex, the male sex drive is stronger due to his potential benefits whereas the potential costs, or at least the lack of benefits, for females means a weaker motivation to engage in sex and the greater plasticity of female sexual behaviours.

We'll keep this in mind as we move on to Ryan and Jethá's next bit of 'evidence'.

They (pp. 272-273) discuss experiments by Meredith Chivers (Chivers et al. 2007) where women and men watched various videos of women and men in varying degrees of possibly erotic activities and their physiological responses were compared to how sexually aroused the subjects reported they felt via a keypad. Ryan and Jethá say to think of this experiment "as an erotic lie detector".

Men, they say, are predictable, and straight men respond to naked women, gay men to naked men, and their conscious and physical arousal match. Women, though, responded physically to everything, including mating bonobos, though they reported that they were not sexually aroused. Their lies, it would seem, are being caught out. Ryan and Jethá then (p. 274) tell the reader of experiments showing that women responded faster to erotic images than any others, as did men. All this, of course, is presented as supporting evidence for the authors' arguments that women's sexual discrimination is caused only by constraints placed on them by society and that women, if society chooses, will rarely say no to any man – or bonobo, it seems.

Ryan and Jethá (p. 272) use a study of sheep and goats (Kendrick et al. 1998) to show that males appear to become fixed to one particular sex object when young whereas females have a "love-the-one-you're-with" approach. Male goats and sheep raised by the other species only reacted to females of that species as sex objects whereas females raised with the other species mated with either. There are a number of problems with this that are not addressed. To start with, offspring are naturally raised by their own species so this is addressing the mechanism which has evolved to avoid hybrid matings. Presumably it is the males who seek out the females so under natural conditions the males are being 'programmed' with the correct species to seek out, and females are not going to be approached by anything other than their own species of male. Where males are the active seekers of matings we would expect selection to act more strongly on them in this respect.

We also need to balance this with the knowledge we have that males of numerous species will attempt to mate with other males, inanimate objects, or other species as noted in Chapter Two. There cannot be many who have not experienced a dog trying to mate with a table leg, or their human leg. For males, producing large amounts of sperm, taking this 'scattergun' approach is better than missing a genuine reproductive opportunity through being overly discriminating (see, for example, Williams 1975).

Females will lose reproductive fitness from poor matings that do lead to conceptions which will normally only involve males *of their own species* so we would expect discriminatory behavior in this *same species* respect. So for females we would expect a 'free choice' to include the avoidance of low-fitness matings when fertile if the situation allows *but* we can very much expect, as Baumeister (and Hrdy) suggest, that female mate choices are made in particular real-world situations comprising complex costs and benefits.

There are as yet no reliable studies on how internet pornography is affecting male 'erotic plasticity' but from what is available and viewed by men it strongly suggests that habituation to initially sought images leads in at least some men to an increased need for novel images which can progress through to sexual activities way beyond their initial preference (Doidge 2007).

So why might human females apparently react sexually to such a varied selection of sexual imagery? Meredith Chivers who carried out these studies is no more believing that women are lying and do want to have sex with bonobos than every other woman. So, what's going on?

Chivers has also looked at research reporting not only genital arousal but also the occasional occurrence of orgasm during sexual assault, and studies showing surges of vaginal blood flow as subjects listen to descriptions of rape scenes. To be logically consistent, Ryan and Jethá would have to conclude that these women are 'lying' about not wanting sex in these situations. Chivers has an alternative explanation which stresses the difference between *reflexive sexual readiness* and desire:

"Ancestral women who did not show an automatic vaginal response to sexual cues may have been more likely to experience injuries during unwanted vaginal penetration that resulted in illness, infertility or even death, and thus would be less likely to have passed on this trait to their offspring".[24]

---

[24] "What Do Women Want?" Daniel Berger January 22, 2009. *New York Times Magazine,* www.nytimes.com/2009/01/25/magazine/25desire-t.html retrieved April 12, 2012

Not exactly the explanation Ryan and Jethá want their readers to think about but one more consistent with female sexuality and their vulnerability amidst males who are bigger, much stronger, and have sex drives that females cannot necessarily avoid. If females only needed to have vaginal lubrication when they actively sought sex with a male of their choice then that is all they would have.

Females have the natural dilution of cervical mucus when the female is ovulating which creates the normal vaginal 'discharge' in primate females that helps to protect her during coitus, along with any vaginal lubrication that is the result of sexual arousal. But what about all the other times when she needs to mate in exchange for food, or to avoid potential infanticide, or to avoid male aggression – situations where she may have to behave proceptively or merely receptively regardless of her interest in copulating? A reflexive sexual readiness would be adaptive, but that does not mean that it is an actual desire for sex *per se,* and it can still mean these sexual encounters would be better avoided if she could gain her benefits without them.

Ryan and Jethá (p. 278) write: "If it's true that women's sexuality is more contextual than most men's, we might need to reconsider a lot of what we think we know about female sexuality". Sarah Hrdy has written about the 'situationally dependent' nature of primate female sexuality (see e.g. Hrdy 1999b). This understanding came from her studies of langur monkeys where she saw females mating outside of oestrus with non-group males and with new males who ousted the resident alpha male from their group: the female's sexual response was adaptive to her because the paternity confusion reduced the likelihood that those males, on becoming a new resident alpha male, would kill her infant.

Female chimpanzees, and to a lesser extent bonobos, mate to appease male aggression. They also mate to get food from males. And, as we saw earlier, female chimpanzees and bonobos engage in sexual activity to improve their social connections with higher status individuals. If female sexual behaviour was simply a matter of the prospect of sexual pleasure then wouldn't we expect to see more sex occurring outside of these situations where the females are obtaining other benefits or avoiding harm?

Ryan and Jethá argue that female access to all the resources they need would mean that females would be more likely to engage in sex, but the very evolution of female sexual receptivity and proceptivity beyond the need to conceive has largely evolved *in order to* access the resources – material and social – the females need.

Ryan and Jethá (p. 278) end their chapter with reference to women's apparent "sexual fluidity" and how women appear to respond more to the person than the person's biological sex due to the desire for emotional intimacy. This conveniently fits with the enjoyment men get from seeing women engaging in sexual activity together but it does not fit with the male's other interest in sex without emotional intimacy. It suggests that women should not be expected to comply with the male's desire for sex without any emotional investment, and that the male's opportunity to engage in 'meaningless sex' will be 'naturally' constrained by the lack of females who are willing to oblige rather than being only due to 'unnatural' social constraints on female sexuality.

Before we get too caught up in the sexual behaviour of modern Westerners we might stop to consider the lack, or rarity, of these behaviours in hunter-gatherers and other non-Western people, and how modern access to pornography, amongst many other novel influences, is shaping the behaviour of modern teenagers of both sexes in ways that did not exist for our pre-agricultural ancestors.

Female sexuality *is* more situationally dependent than male sexuality – and the situations modern females find themselves in today are ones that our ancestors never knew. Our foremothers had no 'singles scene' to deal with, were married around the time of puberty, were mothers by the time they were twenty, had the children to feed and keep alive, and, the evidence strongly suggests, experienced 'vanilla' sex lives.

*

- The discussion by Ryan and Jethá of female sexual frustration and "hysteria" overlooks all positive Western attitudes towards female sexual pleasure and the negative attitudes of many of the partible paternity societies.

- Female copulatory vocalizations are therefore extremely unlikely to be heard, as Ryan and Jethá say they are, in the "upper reaches of the Amazon".

- Copulation calls in chimpanzees and bonobos function to signal social dominance relationships rather than encourage sperm competition. It is untenable to argue that they function to elicit sperm competition in humans.

- The breasts do not "echo" the buttocks as a sexual signal relocating due to bipedalism – buttocks are a *consequence* of bipedalism, and the female fatty deposits on the buttocks evolved even later.

- The authors misrepresent Alan Dixson as saying that the human cervix is similar to the very complex cervix of some macaque species.

- Baumeister, who coined the term "erotic plasticity", notes that while it can be usefully adaptive it also means that the female's greater susceptibility to outside influences can lead to sexual gullibility.

- Female vaginal arousal and lubrication in response to varied depictions of sexual behaviour is more likely a reflexive adaptive response that helps to avoid injury rather than an adaptation to 'enjoy' any and all sexual opportunities.

- This reflexive response is also likely due to an adaptation to be flexible enough to respond positively and competently to sex when she does not particularly want it. This would include the use of sex in exchange for the resources the female does need.

- Most of the situations modern females find themselves in today are ones that our ancestors never knew.

Ryan and Jethá, after their cursory and limited treatment of female sexuality, return again to male sexuality and come to their conclusion about what men want and how women have evolved to accommodate it.

CHAPTER TEN

What Men Want?

Ryan and Jethá begin their CHAPTER TWENTY-ONE discussing paraphilias (abnormal sexual desires and behaviours), how these are mostly found in men, and how they are very hard to change once they are experienced and fixed during a boy's "developmental window". They discuss this to support their view of the rigidness rather than plasticity of male sexual desire and how it cannot – unlike female sexuality – be changed. Perhaps, but if this is so then we would expect there to be some serious questions asked about the novel influences that are shaping the sexual development of young boys today.

The one thing, the authors tell us, male sexual desire does respond to is testosterone. They mention (pp. 281-282) a female-to-male transsexual who describes how testosterone increased his libido so that he subsequently perceived women in such a way that any attractive quality was enough to "flood my mind with aggressive pornographic images just one after another… I felt like a monster a lot of the time. It made me understand men. It made me understand adolescent boys a lot."

No doubt, and we certainly should not hide from or deny what male sexuality can entail, but it would be interesting to discover how this compares to what floods through the minds of hunter-gatherer men. Also we should note that this aggressively sexual male response towards females is much more chimpanzee-like than bonobo.

Ryan and Jethá run through the effects of testosterone on adolescent males and young men, their bodies "screaming for SEX NOW", and how societies try to ignore and to channel this energy elsewhere, which, they say, has been "a centuries-long disaster". Mentioning the suicide rate of young males they suggest the "gut-wrenching, identity-clouding sexual frustration" as a possible cause. They discuss the trouble young people are getting into taking sexual images of under-eighteen-year-old partners or sending sexual images of themselves to "friends".

What they fail to mention is the sending of images without the consent of that person, usually a girl, to multiple others or posting them on the internet. To pretend that sexting and the like is *nothing but* harmless fun and to completely ignore some highly regrettable consequences, including a couple of suicides by young girls, is terribly naïve. As one girl says of what happened when she was thirteen:

"He was 16 and I felt flattered that someone of that age should fancy me. I definitely thought I was in love with him and would have done anything he asked – so when he asked me to send pictures of myself naked, although I felt a bit awkward, I agreed. He sent them on to his friends, the whole school seemed to know about it… the fall-out for me was terrible."[25]

---

[25] 'Sexting: a new teen cyber-bullying 'epidemic'', Glenda Cooper, April 12 2012, *The Telegraph,* http://www.telegraph.co.uk/technology/facebook/9199126/Sexting-a-new-teen-cyber-bullying-epidemic.html retrieved April 13, 2012

This suggests that Baumeister may be right about the potential sexual gullibility of girls and women, possibly due to their 'erotic plasticity'.

There is nothing 'natural' about this modern technology, though the emotions that feed these behaviours have deep roots, including the emotions of sexual jealousy and revenge which can make use of such technology to potentially destroy an ex-girlfriend. Or the technology, along with male attitudes to females and sex, can lead to the everyday 'fun' of their use for the sexual harassment and sexual humiliation of girls. And this all goes on without the protective adult constraints that would naturally have been present in our past.

Children in forager societies often do have some degree of sexual activity with age-mates, but girls are married around the age of puberty while adolescent boys often have to go through many sexless years while they are instructed by their elders in what it takes to become a man and eventually acquire a wife. Are these hunter-gatherer young men screaming for "SEX NOW" and suffering in the same way while their societies channel their energies elsewhere?

A sexless period for adolescent and young adult males is far from unnatural and is also found in other species, including chimpanzees and bonobos. It is probably worth reminding ourselves of a quote earlier from Frans de Waal (1997) with regard to bonobos:

"There is a sharp decline in sexual involvement during a male's adolescence due to the tendency of dominant males to occupy the core of traveling parties where the females are. Only when they enter adulthood and rise in rank do males regain access to receptive females."

It is the norm across many species for males to have to reach a much greater age than females before they can become sexually active, their sex drives blocked by female mate choice and competition from older males.

Ryan and Jethá (p. 285) write that Mangaian youth are encouraged to have sex with one another. As usual, they leave it at that, but we should note that even under this 'encouragement' a

Mangaian girl still only has three or four successive boyfriends between the ages of thirteen and marriage by twenty, while a boy will have ten or more sexual partners before marriage – the boys travel to other islands for other sexual conquests. Mangaian men want more sex than do the women and some husbands will even beat their wife into submission. And it is also worth noting that at the age of thirteen a boy undergoes the superincision ritual when the foreskin is cut longitudinally (Marshall 1971). Ryan and Jethá prefer not to mention this and other similar practices across many cultures, including subincision which is a deep cut along the length of the underside of the penis, choosing only to write negatively (pp. 286-287) about circumcision in the West.

Sex for the Mangaians is focused on the genitals and on thrusting with little concern for foreplay, and the young men's "pride in the pleasure they can provide a woman", noted by Ryan and Jethá, is mostly about the young man's sense of his own manhood and is less of a concern to him once married. There are also rigid rules of modesty for adults and no talk between parents and children about sex. So while the Mangaians allow for some sexual activity between adolescents there are also many aspects of their sexual behaviour we would consider far less 'sex-positive'.

Ryan and Jethá then mention the Muria of central India who had adolescent dormitories (called *ghotuls*) where they were free to sleep together and encouraged to experiment with different partners. The authors obtained this information from Schlegel (1995), so what else does she say?

Schlegel firstly also notes that young males in other primates only mate occasionally with females and do not reach their adult mating frequency until years after puberty. She then writes that human adolescent sexuality is highly constrained even in permissive societies and that girls tend to marry young. In some societies boys engage in homosexual activity as a sexual outlet before becoming heterosexual husbands which brings into question the supposed lack of erotic plasticity or situational dependency in male sexuality.

Of the Muria *ghotuls* Schlegel writes that the girl leader allots sleeping partners and girls are expected to have sex at least some of the time. Those who intend to have sex may retire to a small hut in the compound. In some *ghotuls* couples are paired off and expected to remain faithful though they are unlikely to eventually marry as marriages are arranged, while in others there is a rule against the same couples sleeping together too often.

This practice, Schlegel writes, also has its price. There is jealousy when a favoured partner is paired off with someone else, while the marriage of a beloved to someone else – *ghotul* couples are rarely betrothed to each other – can be emotionally devastating. The Muria also say that they welcome marriage with its privacy and the freedom to express themselves emotionally as well as sexually. So again a little snippet of sexy information from another culture turns out to be not so rosy when we look in a little more detail at that culture's further treatment of sex and marriage.

The Coolidge effect

Ryan and Jethá (p. 288) tell us one variation of the oft-cited story of the visit to a chicken farm by President Coolidge and his wife. The First Lady observes a rooster mating with a number of hens and is told that he is able to mate dozens of times each day so she asks that this be told to the President. On hearing this about the rooster, President Coolidge asks if the rooster is mating with the same hen and is told that it is with different hens so he says to tell *that* to his wife.

Males of many species, as we know, can potentially reproduce with a number of different females at the same time so they will not waste energy mating with females whose eggs they have already fertilized but they will use energy to mate with a female whose eggs they have not yet fertilized. A new female is always potentially a new reproductive opportunity, something Ryan and Jethá (p. 289) only explain as "the invigorating effect of a variety of sexual partners", which is a bit overly positive because for males in many species (from red deer to the little antechinus) this "invigorating" effect in service of the 'selfish genes' in their sperm leads to exhaustion, illness, injury, and even death.

Ryan and Jethá note "that the females of some primate species (including our own) are also intrigued by sexual novelty" though "the underlying mechanism appears to be different for them". True. Selection would not be expected to act on females in this way when they obviously cannot increase their number of offspring, and the potential costs are more likely to reduce their reproductive success. For females the attraction of novelty is more about a better quality mate or avoiding infanticide by males or gaining other important resources (food, protection, avoiding sexual aggression) from the encounter, and a female maximizes her reproductive output by living as long as possible, not by focusing on mating regardless of the long-term consequences to health or offspring survival.

As for chickens, the rooster in his natural environment has to fight off other males to gain his *brief* period of sexual variety. Hens, if they are mated by a low status, undesirable male, are able to shoot his semen right back out (Birkhead 2000), but they also mate willingly with other males besides the alpha male, males who attract them with the food courtship display and whom and the alpha has failed to fight off. Like most polygynous mating systems, the females also mate with different 'alpha males' over time, and many males don't get to mate at all; male domesticated chickens used for egg production don't even get to live past their first day when the chicks are sexed – the other side of the coin of being male is that few are needed when all they have of value is their sperm.

~~~~~~~~~~~~

Ryan and Jethá's main argument (pp. 290-292) focuses on "Phil", whose "perfect life came crashing down" when his wife discovered his affair. The affair had 'invigorated' Phil; even though he 'loved' his wife he says they had become like friends or siblings and that he "wanted to feel alive again" and it "felt like a life-or-death situation".

Ryan and Jethá (p. 292) explain this male hunger for a variety of partners as being due to the Coolidge effect, "men seeking a constant stream of *different* women doing the *same* old things" and that "variety and change are the necessary spice of the sex life of the human male" (emphasis in original). They add that "an intellectual understanding of this aspect of most men's inner reality doesn't make accepting it any easier for women".

In an authors' note to the paperback edition of *Sex at Dawn*, Ryan and Jethá write that some readers had felt this treatment of "philandering Phil" had been biased towards the male. They answer this by saying that men's affairs are more often simply about sexual opportunities whereas women have more complex motivations, are more emotionally motivated, and are less happy with their marriages.

So, men and women are different. True – and therefore they are not compatible or complementary after all. Ryan and Jethá fail to explain how there is going to be all this meaningless sex for men if the women do not want the same thing. Do the men have to lie to these women about their feelings for them or about the status of the relationship?

If we are going to incorporate the differences between the sexes then we need to incorporate the origins of these differences, i.e., the different ways males and females gain reproductive success. With a greater focus on mating rather than parenting, a male can increase his number of offspring especially if one or more of the females do not mate with other males. If a male has fertilized the eggs of one female then he will have a 'selfish gene' interest in looking for other fertile females. The female whom he has already reproductively exploited can be left to focus on parenting his offspring while he puts his own efforts and resources into other mating opportunities rather than those offspring. Good for him (or at least, potentially, his genes). For the reproductively already 'used' mother of his offspring, she has no option but to keep putting her resources into the offspring produced jointly with that male while he diverts his resources elsewhere.

So, how might females similarly 'exploit' males? Having an affair or sexual encounter while fertile is the female threat to the paired male's reproductive success – this is the 'complementary' female behaviour, i.e., what is on the other side of the coin to the male's focus on exploiting multiple females. The female 'side' to Phil's affair would be best seen if we imagine that his wife "Helen" had responded to her husband's affair with the news that she too had expressed her own evolved "nature": she had, in fact, had sex with a variety of men for a variety of reasons, including for their sperm during those periods when their "three brilliant daughters" were conceived. Does an intellectual understanding of the benefits to the 'selfish genes' in a wife make this any easier for a husband to accept? Does he, as Ryan and Jethá's argument suggests he should, have to restrain a 'natural' urge to go and thank those men for 'pitching in'?

In the INTRODUCTION we noted that Ryan and Jethá (in their introduction, p. 2) state that women fare little better than men do in marriage because the wife is left spending her life apologizing for being just one woman. It is surprising that they do not make what would be a much more convincing (for their story) argument that the husband, in turn, spends his life apologizing to his wife for her loss of access to better quality sperm from better quality men for their children – or even, fathers apologizing to their children for being lumbered with his DNA when there is such higher quality available elsewhere.

Just because behaviours evolve due to the way they help, or helped, to spread 'selfish genes' does not make them acceptable. What's more, the success of the 'selfish genes' not only comes at a cost to others but, as we have seen, it is not uncommonly gained at the expense of the health of that body in which those very genes currently reside. The success of genes carries costs – there's nothing more 'natural' than that. Getting through the gateway to the next generation is all that ultimately counts.

But there is worse to come from Ryan and Jethá. They (p. 293) produce their justification for the Coolidge effect with the completely false explanation that "among social mammals...the male drive for sexual variety is evolution's way of avoiding incest"! They say that it evolved to avoid genetic stagnation and to promote genetic diversity in the prehistoric environment, and that when a couple have been together for a long time "they've become *family*" and this "anti-incest mechanism" kicks in! (emphasis in original).

If you believe that you'll believe anything!

There are so many problems with this 'benevolent', almost altruistic, male promiscuity story that it is difficult to know where to begin. Firstly, just like the rooster, if it were not for the interests of other males and the females too, the male would quite happily be the *only* male mating with all the females for his whole lifetime. Behaviours in males have not evolved to promote genetic diversity but to spread as many copies of their own genes as possible. Only the interests of others get in their way.

Secondly, in social animals one sex or the other leaves at puberty and so incest is avoided. When incestuous matings *are* attempted it is males that instigate them and females that try to fight them off. Female reproductive fitness will be more damaged by a pregnancy that results from an incestuous mating because it has removed from the female the opportunity for a better quality pregnancy whereas the male can still mate with other females as soon as he has finished with his sister, daughter, or other close relative.

Thirdly, in chimpanzees, bonobos, and gorillas, it is the females who have the drive to leave the familiar males and seek out novel males so we would expect any stronger motivation against familiarity and towards novelty in our own species to be in females, not males. What's more, our male ape cousins do not stop mating with the females because they have become too familiar; in fact, in chimpanzees it is the older females that are most sexually attractive to the males because they have proved their reproductive capabilities. In bonobos the females rise up the female hierarchy as they age and they presumably become increasingly sexually attractive (have better reproductive prospects) as they do so.

Sexual variety and males taking advantage of sexual opportunities has *nothing* to do with a 'good-for-the-species' incest-avoidance mechanism in evolution; it has *everything* to do with 'selfish genes'.

Ryan and Jethá placed our ancestors in small, dispersed, extended family groups where everyone has sex with everyone else. Their own argument now suggests that the men in these groups would tire of the unavoidably very familiar women. Presumably the male desire for novelty in these groups could only play out by males having sex with each 'novel' female as she reaches puberty. All the men would, presumably, be having sex with all the young women (including, due to unknown paternity, sisters and daughters) until they tired of their familiarity, and the men were no doubt thankful for the continuous stream of pubescent daughters replacing their mothers. Is it possible to get anything *more* incestuous?

Even if our ancestors were in dispersed groups they will have had links with other groups and the movement of young females to new groups at puberty is the most likely original scenario. This movement of females would have created the gene flow throughout the species as it does in our ape cousins. Our links between families and groups were *revealed* by our pair bonds which, for all their problems, were a compromise that was necessary for us on the road to becoming modern humans with our extensive kinship networks.

Testosterone

Ryan and Jethá (pp. 293-294) then talk about higher testosterone levels as if that can *only* mean better health and all things good. They say that it is not only due to the passing of the years that testosterone levels decline but that "monogamy itself seems to drain away a man's testosterone". They correctly note that marriage and fatherhood correlates with lower testosterone levels; their telling value judgment here is that they say the testosterone is being 'drained away' as if this just has to be a bad thing and something to be avoided!

Most men having affairs, they say, are still happy with their marriages while most women are not. Casual contact with attractive women, they argue, has a "tonic" effect on men in that it raises their testosterone level – but they omit to mention that so does playing chess, winning at sport, or watching your team win at sport, and that likely means levels will rise in connection to combat too (Mazur et al. 1992).

Testosterone is a male sex hormone clearly (though not simplistically) linked to male sexual behaviours and their success in out-reproducing the competition, also leading directly and indirectly to greater morbidity and mortality in males compared to females across species. When "Phil" said that his affair felt like a matter of life or death and that he wanted to feel alive again, it makes sense from a 'selfish gene' perspective in that his genes in his sperm were literally seeking life and a future through the gateway of the new female's body (even if that gateway was blocked by the use of contraception). Of course it felt (temporarily) invigorating and essential and out of his control in spite of what part of him knew the consequences would be; would we really expect the feelings that impel the *costly* behaviours that serve the interests of the genes in the sperm to be *easily* consciously overruled?

Ryan and Jethá have been talking about how we deny our non-monogamous 'true nature' but we don't; we all know that men are potentially polygynous, and most of us know and accept that women can sexually desire more than one man, over time if not simultaneously. What we *do* deny is the 'selfish gene' nature of evolution and how the individual organism can be, and will be, sacrificed if it means that more copies of its genes make it through to the future. This is what most people, including Ryan and Jethá, have still failed to grasp. For the authors, evolution is mistakenly thought to be about the good of the species or the group, and about the well-being of the individual organism especially with regard to sexual behaviour. Across species, though, evolved male (and sometimes female) sexual behaviours harm not only others but often the male himself; until we get to grips with this we will not understand why we do what we do, and why doing what comes naturally so often hurts.

Testosterone acts in male sexual competition which includes status competition between males. Interestingly, in humans and other primates the ability of a male to copulate is not under the control of current testosterone levels (only pre-natal levels) which primarily only influence sexual motivation. This, says Kim Wallen (2001), means that social experience and social context can powerfully influence the expression of male sexual behaviour both developmentally and in adulthood. In his studies on rhesus monkeys Wallen found that the sexual behaviour of lower-ranking males is more dependent on the presence of testosterone than is the sexual behaviour of high-ranking males. The lower-ranking males depend on the increase in testosterone levels in order to compete for sexual access to fertile females.

Surbeck *et al.* (2012) looked at testosterone in bonobos. Bonobo males, they note, form dominance hierarchies and compete for access to females, and their mating success varies with rank. But males and females are also codominant and male rank is not only the result of aggression so the relationships between males and females also appear to be important in reproductive success. Because of the detrimental effects of testosterone, such as suppression of the immune system or higher risk of injury, Surbeck *et al.* note that increased testosterone would not be so important if the harmful effects were not offset by increased mating success, i.e., the male body trades this harmful increase in testosterone for the potential benefits to the genes seeking a future in the next generation.

Surbeck *et al.* found that rates of bonobo male aggression and testosterone levels increased in the presence of potentially fertile females, and aggression positively correlated with male rank. But only low-ranking, less aggressive males showed the predicted increase in testosterone levels, and high-ranking males actually exhibited *lower* testosterone levels than low-ranking males during times of mate competition. The stronger relationships between high-ranking males and unrelated adult females meant that these males had lower testosterone levels and did not respond with increased levels during times of sexual competition for mates.

The authors of this study suggest that it is the amicable relations with females that reduce the testosterone levels of the high-ranking males (therefore avoiding the costs of an increase in testosterone). As well as high testosterone levels being incompatible with these intersexual relationships they are also incompatible with paternal care in species that have paternal care (Wingfield *et al.* 1990). Surbeck *et al.* note the similar association between amicable intersexual relations and low testosterone levels in humans, and that men who are involved in romantic relationships have lower testosterone levels and lower testosterone responses to the presence of a woman than do men who are not involved in a relationship with a woman (Burnham *et al.* 2003, van der Meij *et al.* 2008).

None of this suggests a "tonic" effect of increased testosterone levels, it only suggests that increases in testosterone are associated with sexual competition in males but this increase and its detrimental effects are avoided in some males such as those who are in amicable relationships with females and/or are involved in paternal care of offspring. Rather than testosterone being *drained* from males in the negative way asserted by Ryan and Jethá, for bonobos, for intersexual relations, and for male parental care, lower testosterone levels are a benefit to females, to offspring, and to the male himself.

Ryan and Jethá (p. 294) tell the reader about anthropologist William Davenport who lived among Melanesian islanders in the 1960s. Until colonial laws stopped the practice, they tell us that these people avoided monotony by allowing men to have young lovers. Wives were not jealous of these concubines but regarded them as status symbols. So let's have a bit more detail from Davenport (1965) here.

The acquisition of concubines was only open to older, well-established men. The cost of the concubines, who were imported from another island, was ten times the amount paid for a wife and beyond what most individuals could afford so a group of five to ten men of the same 'men's house' would buy one collectively, and one of the men had authority over the concubine and how she was shared. The concubine's social status was really that of a domestic slave: she had no authority over any children she bore and she could be legally sold or even killed by her principal owner. This, Davenport says, is why the brideprice was so high. A normal brideprice signified only partial transfer of rights over a wife from her kin to her husband's kin, but the purchase of a concubine meant absolute severance of all her family relations and she was even given a new name from a stock of personal names reserved for concubines.

The concubine was also a prostitute and sometimes could represent a sound investment as she could be sold on a night-by-night basis to men from other men's houses. Sometimes she might even be sold to another men's house for a profit, or she might just be sold on when the men tired of her and they would then begin to negotiate for a new one. As they grew old, concubines were sold to less prosperous districts and finally given as wives for a normal brideprice. Sometimes concubines served the purpose of a sexual outlet for unmarried sons to prevent them having affairs with single or married women which would lead to serious legal tangles.

This (yet again!) paints a very different picture from the edited version presented by Ryan and Jethá. They are right that these men miss their concubines. As Davenport says, they miss having "women over whom they could exercise absolute authority; on whom they could heap material favors – young women who could remain uncalloused by hard work, and whom they could possess sexually, yet cast off when passion declined or when an opportunity for profit appeared".

Then we are told that Masters and Johnson reported that an ageing male's loss of interest with his partner is due to the monotony of being with the same female and that he can get his interest back if he has sex with a *younger* lover. Are any of us surprised that a male will rise to the occasion when his genes have the prospect of new reproductive outlet to exploit?

These stories and references from Ryan and Jethá show that they are clearly building strong foundations for a conclusion they do not state outright, i.e., that men have sexual preferences for multiple young (fertile) females and are still driven by the most ancient of male traits to attempt to fertilize and exploit as many eggs as possible. Are we surprised that selfish genes that are in male bodies might want nothing less? What Ryan and Jethá miss is that there are also 'selfish genes' in every other organism that, given the choice and the ability, produce *their* behaviours that will act to block these polygynous and exploitative interests.

Ryan and Jethá (p. 295) again give a brief nod towards females also finding sex with the same partner monotonous with time. The sex differences and the reasons women are more likely to look elsewhere when dissatisfied with a partner are not explored and there is only the emphasis on the male 'need' for multiple young females which is hardly news to anyone.

So what is the answer? We are told that modern couples aren't as flexible as the Melanesians and many of the societies surveyed earlier. Now that readers have more details, perhaps we have some insight into this lack of 'flexibility'. Perhaps modern people don't like the whole package that comes with such male dominance and control? Perhaps, looking at the Melanesians, modern Western women don't think men should be able to buy, pimp out, and discard young women that way? Perhaps, looking at some of the other options, modern women would not agree to being 'shared' sexually so the men can form their strong warrior bonds?

Next, the poorly concealed male bias comes even more to the fore when the authors bring in what they say was an argument put forward by sociologist Jessie Bernard in the 1970s that "increasing men's opportunities for sexually novel partners was one of the most important social changes required in Western societies to promote marital happiness". This is quite some spin on what Bernard actually wrote in *The Future of Marriage* (1972/1982)!

What Bernard was, in fact, writing about was that men benefited far more from marriage than did women who would change from happy singletons to depressed wives and mothers with a multitude of psychological problems. Though men for centuries had railed against marriage it turned out to be very good for their psychological health and the majority of their needs. Men only had two complaints: their economic responsibilities and the constraints on their polygynous desires. The former, Bernard says, was being improved by women increasingly sharing this burden, and the latter by what appeared to be a greater tolerance for extra-marital affairs, though she noted that extramarital affairs had historically been tolerated in men anyway, so it was not new. Marriage for men was pretty much sorted.

The big problem was to make marriage better for women. This was Bernard's concern and what her book was about. Women suffered due to their low status as wives, their social isolation, and the burden of all the household chores and of having to minister to all the needs of a husband.

Bernard notes how people often want contradictory things: security *and* freedom, thrills and excitement *and* a safe haven. And this applies also to marriage. She saw young people at the time as moving more away from security and more towards freedom, conceivably, she says, too much so.

Bernard writes that if extramarital affairs become more acceptable it may mean either that the marriage is so strong that it is not threatened by them or that it is so brittle and insignificant that the partners do not care. If there are differences between the pair in what they want the chances are that it is the wife who will have to make the adjustments just as, she says, they have always had to do (fitting nicely, if less willingly, with Ryan and Jethá's argument for greater female plasticity and accommodation of male 'wants'). In reality, she says, husbands are often shocked if a wife actually has sexual relations outside of the marriage.

Bernard covers some very interesting problems, such as the difficulty of equal, companionable relationships in a marriage being able to sustain the sexual interest of the male who prefers a 'battle of the sexes' with the feelings of conquest and female submission for his sexual performance (the testosterone 'high'?).

Male "prophesies" for marriage and sexual relations, she says, are designed to serve male fantasies and dreams. Male swingers, she notes, falsely imagine themselves as cosmopolitan members of the 'jet set', the beautiful people. In reality they often cannot live up to their own expectations. Male bias is also evident in the *ménage à trois* – a secure wife and an available outside partner look just fine to men but it is not good for either woman. As for group marriage, that's a set-up that just aggravates the problems of the two-person marriage.

So what were women looking for? Not what the "male prophets" wanted who, Bernard says, were lacking in sensitivity and certain of their own broadmindedness and that they knew what women wanted or ought to have. In reaction to men's wants the "female prophets" were rejecting men as husbands, and rejecting, even belittling, sex itself. Women wanted better sex, not more. They were rejecting the male model for sexual behaviour and railing against the way it had become mandatory for women to be sexy just as it had been mandatory for them to be frigid a hundred years before; the new norm was no better than the old.

Women were finding sexual 'freedom' to be exploitative of women, and they felt they were still engaging in sex to please men, not themselves. Relationships were still defined by men and only gradually, Bernard says, did it dawn on women that this was a path to destruction.

The "female prophets" wanted meaningful relationships according to Bernard. But mainly they wanted shared roles: shared childcare and household chores. They wanted identities beyond that of a wife, and a greater emphasis on the care of children rather than sex.

Changing the wife's marriage was more important than changing the husband's.

Bernard forecast that marriage would be more about options but she pointed out that options bring conflict and the constant questioning of whether a different option might have been better. There was no utopia to be discovered, and there was also, she said, the question of the pendulum swinging too far in the direction of freedom. People, she noted, shake their fists at the restraints they need and know they must have; they want incompatible things like freedom *and* security but cannot have it both ways, and marriage would always have to be a compromise.

I have written at some length on this book from forty years ago in part because it is so fascinating but also because readers need to compare what Bernard actually wrote with the totally unacceptable "increasing men's opportunities for sexually novel partners was one of the most important social changes required in Western societies to promote marital happiness" distortion by Ryan and Jethá. Jessie Bernard must be turning in her grave.

Are Ryan and Jethá, as we draw to the end of *Sex at Dawn* with their emphasis on a man's need to to keep his wife and his own children while having a succession of novel young lovers, ultimately any different from our Victorian forefathers? It is not news that men are often happy to have sex with young females if the opportunity arises. What they don't want is to lose their wife and children and homes and all those benefits that men gain from marriage. Victorian men could have it all because women could do nothing about it.

When the authors argue that women who have affairs do so because they are unhappy with their husband then it suggests that if the woman has the affair then a divorce would be the right option for the husband to choose. But men who have affairs, they say, still 'love' their wife so Ryan and Jethá argue that *they* should not be punished by a divorce nor, it would seem, do they need to do anything that could improve their marriage.

If the authors' is a genuine argument for a change in men's attitudes towards female sexuality and paternity rather than simply a return to the times when men were able to have extramarital sex without the threat of divorce, then surely the emphasis and the culmination of their book would be about achieving this; it would be about convincing the men to accept a genuinely novel way of being rather than telling women how the old male privileges need to be reinstated. Otherwise, the only difference between Ryan and Jethá's concluding arguments and Victorian times is that our Victorian foremothers could not divorce a husband even if they wanted to do so.

Hunter-gatherer women can and do use divorce if their husband seeks sex with others, and the interdependency of spouses and the equal right to divorce helps to constrain some of the reproductively selfish behaviours. Ryan and Jethá's emphasis on the removal of that constraint on husbands simply puts us back to the male privileges that dominated marriage over the last 10,000 years.

Ryan and Jethá at no point throughout *Sex at Dawn* discuss the male need for a pair bond and when, where, or why it evolved. Their whole book is about our supposedly bonobo-like recent ancestry, so why do so many men marry and gain such health and well-being benefits from marriage? Why do men marry at all? Why did successful and attractive "Phil" with no shortage of sexual opportunities *care* about having a wife in the first place, never mind losing her?

If the book concluded with more about how men should be letting their wives conceive 'their' children while mating with multiple men, and how they should not have any qualms about raising the children of these men, then that would really be something revolutionary. What Ryan and Jethá offer is ultimately no revolution at all, only a primary argument for a return to marriages where men never need to pass up a sexual opportunity for fear of the threat of losing a wife and family. As Jessie Bernard says about "male prophets" who write on such things, it is male fantasy and male dreams where men get to have their cake and eat it too.

Ryan and Jethá next (p. 296) take us again to that wonder drug of all wonder drugs, testosterone, and how the thrill of a novel young sex partner can produce a 'hit', if only temporarily. When the thrill is over, they say, many men realize that the abandoned partner was not so bad after all, with all kinds of benefits he actually needed (apart from those 'reproductive' ones for his selfish genes which they have failed to acknowledge). This is where we are told we need to make smarter decisions about our long-term relationships and how "this understanding requires us to face some uncomfortable facts".

We are told that when "Phil" felt he faced a life-or-death decision he may really have been doing so because the boost to his testosterone level would be protecting his long-term health. Really? They present evidence from a few studies of the association between the decline of testosterone with age and various illnesses. Though they do not mention it, women too have health problems when they age such as the loss of the protective influence of the female hormones on the heart and the bones. Whereas older women sometimes opt to take hormone replacements the implied remedy for the older men is sex with young women. Are we really meant to swallow that? Probably, if we recall that in this fantasy world of *Sex at Dawn* young females are not meant to make mating decisions with their heads – or eyes, it would seem – but let all the men in, young and old, ugly or handsome, and let those wonderful sperm fight it out.

Ryan and Jethá (p. 298) come to their conclusion:

"If it's true that most men are constituted by millions of years of evolution, to need occasional novel partners to maintain an active and vital sexuality throughout their lives, then what are we saying to men when we demand lifetime monogamy?" Most men, they say, don't realize the conflict between the demands of society and their own biology until they have been married for years and have joint children and joint property.

We should note that it is surely no coincidence that this realization hits men when they have already obtained the lifetime commitment of one woman to what are his children while he still has plenty of sperm – and selfish genes – ready and waiting to exploit younger women not yet used up reproductively. Having maxed out their wife then it is only reproductive logic to look for openings elsewhere.

The man's options, Ryan and Jethá say, are to lie and try not to get caught, give up on sex with anyone other than his wife and perhaps resort to porn and Prozac, or divorce and start over again.

No option is suggested to gain some understanding of the nature of sexual reproduction and how this is really driven by the 'selfish' genes seeking as many eggs as possible to get them into the next generation; selfish genes in males that really could not care less about his long-term health and well-being. For all the talking-up of testosterone it has evolved to *serve* the 'selfish' genes in sexual competition with all its costs, and certainly not to serve the long-term health and interests of the body in which they have their temporary home.

The options for women (p. 300) are to pretend you do not know your husband is cheating, go out and have your own revenge affair even if you don't want one, or destroy your family by calling in the lawyers (note that only the wife, not the husband, is destroying the family by their chosen actions).

Ryan and Jethá say that the use of the term 'cheating' "echoes the standard narrative of human sexuality in its implication that marriage is a game one player can win at the expense of the other". They now briefly mention women who 'trick' a man into supporting children he thinks are his, and the "baby-daddy" who manages to impregnate a string of women who raise his children while he moves on to his next conquest. They say that in any true partnership "cheating cannot lead to any sort of victory. It's win-win or everybody loses".

The main aim of Ryan and Jethá's argument has been to remove any reason for a 'natural' conflict of interests between the sexes so this has to include the removal of 'cheating' – not the *behaviour* we call cheating, just the way we react to that behaviour. They have focused on the husband/father's 'need' for extra-marital sex but this is nothing new and is hardly distinguishable from the historical victory that has been had by married men. How much more believable their intentions would be if they had focused their conclusions on things that *men* find hard to face within their reproductive pair bonds, i.e., wives having sex with and becoming pregnant by other men.

There is a strong likelihood that rather than men 'naturally' not being disposed to care whether the children they raise are their own or not, this 'freedom' would instead lead to a lot more 'baby-daddies' – men focused purely on mating with no interest at all in parenting. And this brings us again to that great unanswered question in *Sex at Dawn:* where on earth did those 'partnerships' – 'true' or not – between men and women come from? Why do they exist? How and why did they evolve? Ryan and Jethá completely fail to address this question.

There is indeed a lot that rightly makes us uncomfortable about our and other species' naturally selected sexual self-interests but if we have to feel uncomfortable we should at least let it be in response to some facts.

~~~~~~~~~~~~~

The final chapter heading, *Confronting the Sky Together*, suggests that women and men, thanks to what the authors have revealed to us, will be able to reach a better understanding. Ryan and Jethá *again* use a quote from E. O. Wilson (1978) about the costs of going against our nature. As noted a number of times already, Wilson believed that the pair bond and concern about the sexual fidelity of a spouse *was* part of our nature and he was only attacking the Catholic Church and their attitudes towards birth control and homosexuality.

Ryan and Jethá start by suggesting that most of us take sex too seriously and that it's "just sex".

Without an understanding of the evolution of sex and the sexes this seems like a valid point. But is the rubbing of bodies and body parts together until one person ejaculates hundreds of millions of gene packets into the other something evolution came up with 'just' for the fun of it? Did sexual selection 'just' happen to provide the male with the biggest neurochemical reward for this behaviour because sex is "just sex"?

Ryan and Jethá write that sex is essential to survival like food and water but that is not strictly true as sex is only essential to the survival of genes through time, and people who choose celibacy are not noticeably dropping dead in the street. They argue that sexual satisfaction should be more easily available in order to make it less problematic but what they really mean is that sex with attractive young females should be more available to all men. They say that we appear to be flowing back towards hunter-gatherer sexual casualness, but sex in hunter-gatherers is anything but casual, as we have seen.

Again they mention the Siriono who, they say, rarely if ever lack for sexual partners and where sex anxiety seems to be remarkably low. Recall that the Siriono live in small inbreeding groups under one roof and in a constant state of hunger, which makes food anxiety their main experience. Men use food to get sex, women are subservient, marriage is monogamous or polygynous, and women receive the blame for all adultery which is hidden as much as possible. A man alone with a woman in the forest may throw her to the ground and have sex with her without so much as a word. Sex is generally a violent and rapid affair, and while kissing is unknown, biting occurs.

How would it feel to live in such a world, Ryan and Jethá ask. Hungry, probably, and not that great for women. Also, living with the same few dozen people under the same roof with only a small subset of potential sex partners, we would expect, according to Ryan and Jethá, for familiarity to dampen all sexual excitement and interest.

Ryan and Jethá (p. 304) write that "[t]he false expectations we hold about ourselves, each other, and human sexuality do us serious, lasting harm". But more than that (and what they refuse to acknowledge) the lack of understanding of the 'selfishness' at the root of sexual reproduction does us the greatest harm, and it even allows for some people to impose their own agendas under the cover of revealing the truth to those who know no better.

Of course, talking about and understanding the selfish nature of evolved sexual reproduction and the inherent conflicts between the sexes are not easy conversations to have with a reproductive partner. But understanding how desires and behaviours have their roots in the interests of 'selfish genes' producing self-interested traits that have evolved over hundreds of millions of years can help us to look at them with greater objectivity rather than having an automatic and obedient behavioural response in service to eggs or sperm.

Simone de Beauvoir in *The Second Sex* lamented how "the woman is adapted to the needs of the egg rather than to her own requirements". We did not then understand how males too are adapted to the needs of *their* gametes rather than their own requirements. Men can more easily feel rewarded for serving the needs of their sperm; rewarded time and time again for every successful ejaculation. The 'natural' downside for men and males in general is competition with other males, injury, possible exclusion from *any* mating opportunities, more illness than females, and shorter lifespans.

The natural negatives have been screened out by Ryan and Jethá to leave only multiple acts of 'just sex' to stand alone: complete service to the sperm without acknowledgment of a single, natural negative consequence. For women, of course, it is different. Female bodies have to go through a lot after copulation for them to have experienced reproductive success; the female body cannot be fooled that it is reproducing copies of its genes as the male body can. We are the descendants only of females who did reproduce and their success did not come about by being pseudo-males.

Ryan and Jethá say that one difficulty women have is in understanding how men can dissociate sexual pleasure from emotional intimacy. Yet their pre-history story is one where sex is *only* between people who are emotionally intimate so it would have been useful for them to explain the roots of this dissociation. The second thing they need to explain is: if women do not want 'just sex', who are the men going to have their meaningless sexual encounters with?

Does female erotic plasticity mean that we can, and should, 'educate' all girls to dissociate sexual pleasure from emotional intimacy in order to accommodate this male desire? Would this be *natural* female sexual behaviour? Are we *really* going to learn what female sexuality is or should be from what some *men want* it to be?

We also need to think more about what 'emotional intimacy' actually means, and why it might matter to women but not men. Might it not have something to do with the different impact a partner's quality has on the reproductive fitness of the two sexes? Might it not have something to do with which direction genes travel during sex and the different consequences for the two sexes? While sexual intercourse is *the goal* for genes in sperm it is certainly not the goal for genes in eggs.

Might it not also have something to do with a potential fate that only females suffer: getting pregnant by the *wrong* male? Might we not expect females to do better (to have greater evolutionary 'fitness') if they gain direct (food, protection, etc.) and/or indirect (genetic) benefits from sex rather than act in a pseudo-male, 'indifferent about what happens next' way? And it can only be *pseudo*-male behaviour because women are never going to be shooting their eggs into the male; nor will the use of contraception change evolved female sexuality into evolved male sexuality.

Female sexuality *is* far more complex than that of the male. Sperm-serving traits have not altered as much over hundreds of millions of years as egg-serving traits have done. And while there is plenty we still do not understand about our evolved sexual and reproductive behaviours, we at least can be sure that the portrayal of our ancestral sexual behaviours in *Sex at Dawn* is not supported by evidence.

Whatever people choose to do in the modern world is up to them. But the last thing we need is another false story about our ancestors and about the benign nature of sex and sexual reproduction.

As we come to the end of this story, Ryan and Jethá (p.307-308) tell us that the first American swingers were World War II air force pilots and their wives, "like warriors everywhere, these 'top guns' developed strong bonds with one another" and those who survived would be understood to then look after the widows. How much truth there is in this is impossible to say but any suggestion that this is our ancestral behaviour does not even fit with the authors' own argument that our ancestors were *not* involved in warfare so this kind of male bonding – which exists in chimpanzees but not bonobos – should not have been necessary.

What we do know is that hunter-gatherers, like every other society, do not have a casual attitude towards sex and reproduction. The nuclear family with serial monogamy, if not life-long monogamy, is found everywhere and sexual fidelity is expected if not always achieved. Nowhere is extramarital sex open and without at least some significant negative consequences.

*

- In forager societies girls are married around the age of puberty while boys often have to go through many sexless years while they are instructed by their elders in what it takes to become a man and eventually acquire a wife. Adolescent and young adult males, like males of other species, 'naturally' have their sexual interests curtailed.

- Though the Mangaians allow for some sexual activity between adolescents there are also many aspects of their sexual behaviour we would certainly consider far less 'sex-positive'.

- Though the Muria also allowed for sexual activity for adolescents this too had its problems for those involved, and the Muria welcomed marriage with its privacy and the freedom to express themselves emotionally as well as sexually.

- The "invigorating" Coolidge effect in service of the 'selfish genes' in their sperm also leads males of many species to exhaustion, illness, and even death, as well as running in tandem with male contest behaviours which exclude many or most males from mating at all.

- The Coolidge effect has *nothing* to do with an incest-avoidance mechanism in evolution; it has *everything* to do with reproductive self-interest.

- Rather than testosterone being *drained* from sexually monogamous males in the negative way asserted by Ryan and Jethá, for bonobos, for intersexual relations, and for male parental care, lower testosterone levels are a benefit to females, offspring, and the male himself.

- Further details from William Davenport's study of a population of Melanesian islanders shows us a *very* different picture of male sexuality and the role of concubines than the one that is presented by Ryan and Jethá.

- Jessie Bernard was *not* arguing for an increase in men's opportunities for sex with novel partners in order to promote marital happiness. She wrote that "male prophets" who write on such things produce male fantasies and male dreams where men get to have their cake and eat it too.

- *Sex at Dawn* concludes, as it began, with the focus on extra-marital sex for husbands, now with numerous false arguments and justifications for why there should be no objections or comeback from a wife.

- If we have to feel uncomfortable about the differences between the sexes we should at least let it be in response to facts rather than fantasy.

- Ignoring the 'selfish gene' nature of evolution will forever leave us floundering in the dark.

Sluts or Whores?

My lengthy response to *Sex at Dawn* was not inspired by any interest in prescribing what anyone should or shouldn't do today. Ryan and Jethá say that they don't know what readers should do with the information they have provided, so my only concern is that readers should at least be aware that the information they provide is clearly very seriously flawed.

There is, of course, so much more that could be looked at, such as competition between femaies and mate-guarding by females, male mate choice, individual variation, and flexibility. But my primary aim has only been to demonstrate how the authors have distorted snippets of information from generally reliable sources to produce a fiction dressed as fact backed by science.

In addition to many serious errors there are numerous omissions of what is highly relevant information if we are not to be left with a misleadingly fictitious picture of sexual behaviours in other societies. And though my added information about natural and sexual selection and other species may be of limited interest to the readers of *Sex at Dawn*, understanding evolution as a whole is crucial to all of us if we are to understand ourselves better and to grasp the evolutionary pressures on our sexual relationships.

The virtual removal by Ryan and Jethá of pre-copulatory female mate choice in our ancestors is particularly disturbing, and even more so when this is added to their 'evidence' that women's *bodies* want a lot more sex than their *minds* want. A belief in a repressed natural female desire for sex with all-comers, and the potential removal of a woman's right to have her "no" taken seriously, is obviously a serious, and potentially very dangerous, error. But for Ryan and Jethá anything that leads to more sex can only be good.

Some readers may still wish to argue that females are not being denied the opportunity to say "no", but the actual arguments throughout *Sex at Dawn* are very much about females rarely, if ever, having any reason to refuse sex. As soon as we do incorporate "no" and include *any* (male or female) sexual preferences, as will inevitably happen in the real world, we are then faced with sexual rejection and jealousies and competition which the authors deny existed in our prehistory.

Connected to this is their argument that 'slut' is good and 'whore' is bad, yet females across species are 'whores', not 'sluts'. Rather than an ejaculation, i.e., the potential *end* point of a male's reproductive effort, reproductive success for females is about translating actual resources into actual offspring. Resources can be food but also they include protection for the female and her young or simply avoiding harm to herself or her young. Our female ancestors have not been sluts but they have been whores.

Exchanging sex for food or other resources when not actually fertile is still female *reproductive* behaviour in that it still contributes to her reproductive fitness – just as female langurs mating with males though the females are not actually fertile is *reproductive* sex because it concerns avoiding infanticide by those males. The direct benefits females receive in exchange for sex are essential to females of many species, and this 'whoredom' is totally natural, including for bonobo females. Pseudo-male 'slut' behaviour is not a *better* female sexual behaviour – though, of course, 'free' sex is always better for a male.

Females may also mate when they are not actually fertile as a way of competing with other females, as a study of pregnant western gorillas suggests (Doran-Sheehy, Fernandez, and Borries 2009). Rather than the gorillas mating only when ovulating, these gorilla females were soliciting sex from the male when they were already pregnant, and the silverback mated preferentially with females on the basis of rank rather than reproductive state.

The pregnant female gorillas mated more often on days when other females were mating with the male but stopped mating altogether if there were no other females mating with him. Females harass other females in the context of mating as well as in other contexts, and low-ranking females produce fewer offspring, so this female-female competition for higher rank and greater attention from the male appears to be a female reproductive strategy. The authors of this gorilla study suggest that female-female competition should be considered as a potential factor in the evolution of nonconceptive mating in humans.

Hopefully the expanded information about our ape cousins in Chapter Three has shown how limited and distorted the representations of them are in *Sex at Dawn*. In particular it is essential that we gain a much better understanding of our bonobo and chimpanzee cousins, and put an end to the false representations of bonobo sexual behaviour which does not help us understand our own.

The expansion of information on partible paternity societies in Chapter Four reveals a glaring disparity between what actually occurs and Ryan and Jethá's stories, as does the expanded information on all the other societies they cite, from the !Kung to the Mosuo. While the Mosuo have 'lost' marriage sometime in recent history, marriage otherwise exists everywhere but we get no explanation from Ryan and Jethá as to why this is so, and how such a practice arose in societies never touched by agriculture or settlement.

Ryan and Jethá eliminated extra-pair sexual 'cheating' by removing the existence not only of *sexual* monogamy but also *social* monogamy for our supposedly bonobo-like ancestors. This is obviously incorrect. If they had included the evolution of socially monogamous pair bonds then they would have had to explain how and why such pair bonds evolved in our ancestors and their whole story would have crumbled. So they simply ignore them – at least, that is, until their conclusion where they reinstate marital benefits for husbands alongside their freedom to engage in extra-marital sex, i.e., just what our patriarchal forefathers enjoyed.

Along with the ubiquity of marriage we find jealousy and a concern about paternity everywhere. We also find a sexual division of labour and an interdependency of spouses that is totally ignored by Ryan and Jethá, as is the very different evolutionary path our ancestors took compared to our chimpanzee and bonobo cousins.

The old idea that monogamy, pair bonds, and marriage evolved so that males provided their wife or wives with meat in exchange for her foraged resources is probably wrong because, as is often pointed out, the large game is often shared amongst the group. So the question regarding the role of men provisioning their own wife and family still needs greater understanding. We saw in the Aché that the woman and her children *only* get a share of meat if they have a husband/father/hunter; no man in the family who hunts means there is no meat for that family.

In the Hadza it has been pointed out by researchers that the best hunters share the meat in the group so their wife does not necessarily benefit in this respect from being married to him. Yet women do favour good hunters as husbands and, as we have seen across many societies, women will exchange sex for gifts of meat that they otherwise would not get. One study suggests that in the Hadza it is honey which is important in the husband-wife relationship: honey is not widely shared but given by a man to his wife, and this can be a very important resource for lactating mothers who are less able to forage at this time (Marlowe 2003).

Most Hadza are serial monogamists, though lifelong monogamy, as we saw earlier, is not uncommon. Women are frequently the ones to initiate a break-up and a main reason for this, along with infidelity, is that the husband is an incompetent hunter. Michael Finkel (2009) writes about the Hadza group he stayed with where one man was 30 years of age and not married "bedevilled, perhaps by the five-baboon rule": a man cannot get a wife until he has successfully hunted five baboons. The standard !Kung explanation for why a particular poor hunter remains a bachelor is "women like meat" (Hrdy 1999a).

Another study by Marlowe of the Hadza found that men with more biological children in the camp seemed more motivated to hunt and also more inclined to channel extra meat to children they believed they had actually sired (cited in Hrdy 2009).

It seems quite likely that the social status of a good hunter is something which benefits a wife and her children even if they gain no obvious benefit from extra big-game meat itself. Social rank, as in other social species – including bonobos – will always have its advantages.

The universality of marriage, women exchanging sex for meat and other resources, women preferring good hunters as husbands, husbands and biological fathers channelling resources towards their own offspring – all these exist under the umbrella of sharing and cooperation.

What we certainly do not have is women all shared by the men in a group. Whatever system 'sorts itself out' in the conflicting reproductive interests of the men and the women, what is clear is that for women all men are certainly not sexually welcome, all women are not equally attractive to the men, and for men there is a willingness to buy sex alongside a more direct and constant investment in children where the likelihood of paternity is the greatest.

Though warfare was unlikely to have been common in our ancestors, the leisurely, intentionally hungry, and intentionally controlled population size argued for by Ryan and Jethá does not hold up to scrutiny. Intentionally calorie-restricted, infertile women certainly does not fit with the 'Venus' figurines nor the authors' own arguments for the pleasures of curvaceous, sexy, fertile ancestral women.

The arguments attempted by Ryan and Jethá in order to have readers believe that human testes have only shrunk from bonobo size in the last ten thousand years are extremely weak and some, such as the *seasonal* change in some species, are again just plain wrong. Their presumptions about the inherent benefits of sperm competition and testosterone also do not stand up to scrutiny.

Even worse, the treatment of female bodies and sexual responses is shallow and sometimes even potentially harmful as noted above. At the same time, looking into the information on female copulatory vocalizations at least provided the opportunity to discover some very interesting studies on the way these function to signal social dominance relationships in chimpanzees and bonobos, while female 'erotic plasticity' raised some important questions about the potential sexual gullibility of women and girls.

Finally Ryan and Jethá came to their concluding chapters and their argument for the extramarital sexual encounters of men to be freely accepted. We were incorrectly told how these are virtually essential for a man's health and that marriage and children harmfully drain a man's testosterone. Again we had the edited versions of sexual behaviour in other cultures, including that of Davenport's study of the use of concubines by a population of Melanesian islanders which was misrepresented in a purely positive light.

We also were told a false incest-avoidance explanation for the Coolidge effect, and even that Jessie Bernard's arguments about improving marriage were in support of men's desires for novel sexual partners.

It is significant that in the book's conclusions the authors' earlier arguments for an insatiable female libido disappeared and became one of women wanting emotional sexual attachments and wives not seeking sex with other men if they were happily married. There was no explanation of where the husband's partners for their 'meaningless' and casual sexual encounters would come from. The supposed compatibility of the two sexes suddenly was gone and we were left with husbands keeping their marital benefits while the wives were admonished for not allowing them their extra-marital sex too. Hunter-gatherer women would not accept a deal like that, though our Victorian foremothers, and perhaps all our foremothers of the past ten thousand years, have not had that choice.

The fundamental problem we have with sex is that we imagine that it is *meant* to be fun and easy – 'naturally' about all things good. When it feels bad we imagine that this must be due to some 'unnatural' influences and constraints, and if we could only rid ourselves of these, everything would be great. But the search for a *naturally selected* answer to our sexual and relationship frustrations is only ever going to be a fruitless quest, as a look at the naturally selected sexual traits of other species so clearly shows.

The best we can hope to do is to gain a greater understanding of the evolution of sex and sexual reproduction on which to base our choices. That understanding is not to be found in *Sex at Dawn*.

REFERENCES

Ales, C. (2002). A Story of Unspontaneous Generation: Yanomami Male Co-Procreation and the Theory of Substances. In S. Beckerman and P. Valentine (Eds.), *Cultures of Multiple Fathers: The Theory and Practice of Partible Paternity in Lowland South America.* Gainesville, FL: University Press of Florida.

Alexander, R. D., and Noonan, K. M. (1979). Concealment of ovulation, parental care, and human social evolution. In N. A. Chagnon, and J. W. Irons (Eds.), *Evolutionary Biology and Human Social Behavior.* North Scituate, MA: Duxbury Press.

Amaral, I. Q. (2008). Mechanical analysis of infant carrying in hominoids. *Naturwissenschaften,* 95(4): 281–292.

Andersson, M. (1994). *Sexual Selection.* Princeton, NJ: Princeton University Press.

Andrade, M. C. B. (1996). Sexual Selection for Male Sacrifice in the Australian Redback Spider. *Science,* 271(5245): 70-72.

Angier, N. (2001). A Fresh Look at the Straying Ways of the Female Chimp. *New York Times,* May 15, 2001.

Arnqvist, G. (1997). The evolution of animal genitalia: distinguishing between hypotheses by single species studies. *Biological Journal of the Linnean Society,* 60: 365–379.

Arnqvist, G., and Rowe, L. (2005). *Sexual Conflict.* Princeton, NJ: Princeton University Press.

Barash, D. P., and Lipton, J. E. (2001). *The Myth of Monogamy: Fidelity and infidelity in animals and people.* New York: Henry Holt.

Barton, R. A. (1999). Socioecology of baboons: The interaction of male and female strategies. In P. M. Kappeler (Ed.) *Primate Males: Causes and Consequences of Variation and Group Composition.* Cambridge: Cambridge University Press.

Bateman, A. J. (1948). Intra-sexual selection in Drosophila. *Heredity,* 2: 348-368.

Baumeister, R. F. (2000). Gender differences in erotic plasticity: The female sex drive as socially flexible and responsive. *Psychological Bulletin,* 126(3): 347-374.

Baumeister, R. F. (2004). Gender and erotic plasticity: sociocultural influences on the sex drive. *Sexual and Relationship Therapy,* 19(2): 1468-1479.

Beckerman, S., Erickson, P. I., Yost, J., Regalado, J., Jaramillo, L., Sparks, C., Iromenga, M., and Long, K. (2009). Life histories, blood revenge, and reproductive success among the Waorani of Ecuador. *PNAS,* 106(20): 8134-8139.

Beckerman S., and Lizarralde, R. (1995). State-tribal warfare and male-biased casualties among the Bari. *Current Anthropology,* 36(3): 497-500.

Beckerman, S., et al. (2002). The Bari Partible Paternity Project, Phase One. In S. Beckerman and P. Valentine (Eds.), *Cultures of Multiple Fathers: The Theory and Practice of Partible Paternity in Lowland South America.* Gainesville, FL: University Press of Florida.

Beckerman, S., and Valentine, P. (Eds.). (2002). *Cultures of Multiple Fathers: The Theory and Practice of Partible Paternity in Lowland South America.* Gainesville, FL: University Press of Florida.

Bedhomme, S., Prasad, N. G., Jiang, P.-P., and Chippindale, A. K. (2008). Reproductive behavior evolves rapidly when intralocus sexual conflict is removed. *PLoS One,* 3(5): e2187.

Bernard, J. (1972/1982). *The Future of Marriage.* New Haven: Yale University Press.

Birkhead, T. R. (2000). *Promiscuity: an evolutionary history of sperm competition.* Cambridge, MA: Harvard University Press.

Blum, D. (1998). *Sex on the Brain: The Biological Differences Between Men and Women.* London: Penguin.

Boehm, C. H. (1999). *Hierarchy in the Forest: The Evolution of Egalitarian Behavior.* Cambridge, MA: Harvard University Press.

Boesch, C. and Boesch-Achermann, H. (2000). *The Chimpanzees of the Tai Forest.* New York: Oxford University Press.

Boesch, C., Hohmann, G., and Marchant, L. F. (2002). *Behavioural Diversity in Chimpanzees and Bonobos.* Cambridge: Cambridge University Press.

Boesch, C., et al. (2007). Fatal Chimpanzee Attack in Loango National Park, Gabon. *International Journal of Primatology*, 28(5): 1025-1034.

Boesch, C., et al. (2008). Intergroup conflicts among chimpanzees in Tai National Park: Lethal violence and the female perspective. *American Journal of Primatology,* 70: 519–532.

Bonnicksen, A. L. (1991). The Embryo as Patient: New Techniques, New Dilemmas. In J.M. Humber and R. F. Almeder (Eds.), *Bioethics and the Fetus: Medical, Moral and Legal Issues (Biomedical Ethics Reviews).* Totowa, NJ: Humana Press.

Bowden, R. (1992). Art, Architecture, and Collective Representations in a New Guinea Society. In J. Coote and A. Shelton (Eds.). *Anthropology, Art, and Aesthetics.* Oxford: Oxford University Press.

Bowman, E. A. (2008). Why the human penis is larger than in the great apes. *Archives of Sexual Behavior,* 37(3): 361.

Bradley, B. J., et al. (2004). Dispersed Male Networks in Western Gorillas. *Current Biology,* 14: 510-513.

Brennan, P. L. R., Clark, C. J., and R. O. Prum. (2010). Explosive eversion and functional morphology of the waterfowl penis supports sexual conflict in genitalia. *Proc. R. Soc. B.,* 277(1686): 1309-1314.

Bribiescas, R. G. (2001). Reproductive ecology and life history of the human male. Reproductive ecology and life history of the human male. *American Journal of Physical Anthropology,* 116: 148–176.

Burnham, T. C., Chapman, J. F., Gray, P. B., McIntyre, M. H., Lipson, S. F., and Ellison, P. T., (2003). Men in committed, romantic relationships have lower testosterone levels. *Hormones and Behavior,* 44: 119–122.

Carnahan, S. J., and Jensen-Seaman, M. I. (2008). Hominoid seminal protein evolution and ancestral mating behavior. *American Journal of Primatology,* 70: 939-948.

Carter, C. S., and Getz, L. L. (1993). Monogamy and the Prairie Vole. *Scientific American,* 268(6): 100-106.

Chapais, B. (2008). *Primeval Kinship: How Pair-Bonding Gave Birth to Human Society:* Cambridge, MA: Harvard University Press.

Chen, F. C., and Li, W. H. (2001). Genomic Divergences between Humans and Other Hominoids and the Effective Population Size of

the Common Ancestor of Humans and Chimpanzees. *Am J Hum Genet*, 68 (2): 444–456.
Cheng, K. M., Burns, J. T., and McKinney, F. (1983). Forced copulations in captive mallards III. Sperm competition. *Auk*, 100: 302-310.
Chernela, J. M. (2002). Fathering in the northwest Amazon of Brazil. In S. Beckerman and P. Valentine (Eds.), *Cultures of Multiple Fathers: The Theory and Practice of Partible Paternity in Lowland South America.* Gainesville, FL: University Press of Florida.
Chippindale, A. K., Gibson, J. R.,and Rice, W. R. (2001). Negative genetic correlation for adult fitness between sexes reveals ontogenetic conflict in *Drosophila*. *Proc. Natl Acad. Sci. USA*, 98(4): 1671–1675.
Chivers, M. l., Seto, M. C., and Blanchard, R. (2007). Gender and sexual orientation differences in sexual response to the sexual activities versus the gender of actors in sexual films. *Journal of Personality and Social Psychology*, 93: 1108-1121.
Clark, A. G., and Civetta, A. (2000). Protamine wars. *Nature*, 403: 261-263.
Clark, N.L., and Swanson, W. J. (2005). Pervasive adaptive evolution in primate seminal proteins. *PLoS Genet.*, 1(3):e35.
Clark, W. R. (1996). *Sex and the Origins of Death.* Oxford: Oxford University Press.
Clay, Z, and Zuberbühler, K. (2012). Communication during sex among female bonobos: effects of dominance, solicitation and audience. *Scientific Reports* 2: 291.
Copeland, S. R., Sponheimer, M., *et al.* (2011). Strontium isotope evidence for landscape use by early hominins. *Nature*, 474: 76–78.
Crocker, W. H. (2002). Canela "Other Fathers": Partible Paternity and Its Changing Practices. In S. Beckerman and P. Valentine (Eds.), *Cultures of Multiple Fathers: The Theory and Practice of Partible Paternity in Lowland South America.* Gainesville, FL: University Press of Florida.
Crocker, W. H., and Crocker, J. (1994). *The Canela: Bonding though kinship, ritual, and sex.* Fort Worth TX: Harcourt Brace.
Cronin, H. (1991). *The Ant and the Peacock.* Cambridge: Cambridge University Press.
Cronin, H. (2006). The Battle of the Sexes Revisited. In A. Grafen and M. Ridley (Eds.), *Richard Dawkins: How a Scientist Changed the way we Think.* Oxford: Oxford University Press.

Darwin, C. (1871) *The Descent of Man*. Retrieved July 7, 2011 from http://www.darwin-literature.com/The_Descent_Of_Man

Davenport, W. H. (1965). Sexual patterns and their regulation in a society of the southwest Pacific. In F. A. Beach (Ed.), *Sex and behavior*. New York: John Wiley & Sons.

Dawkins, R. (1976). *The Selfish Gene*. Oxford: Oxford University Press.

de Beauvoir, S. (1953/1981). *The Second Sex*. Harmondsworth, England: Penguin Books.

de Waal, F. B. M. (1989). Behavioral Contrasts between Bonobo and Chimpanzee. In P. G. Heltne and L. A. Marquardt (Eds.), *Understanding Chimpanzees*. Cambridge, MA: Harvard University Press.

de Waal, F. B. M. (1995). Sex as an Alternative to Aggression in the Bonobo. In P. R. Abramson, and S. D. Pinkerton (Eds.), *Sexual Nature Sexual Culture*. Chicago: University of Chicago Press.

de Waal, F. B. M. (2005). *Our Inner Ape: The Best and Worst of Human Nature*. London: Granta Books.

de Waal, F. B. M. (2007). Bonobos, Left & Right Primate Politics Primate Heats Up Again as Liberals & Conservatives Spindoctor Science. *eSkeptic*. Retrieved July 4, 2011 from http://www.skeptic.com/eskeptic/07-08-08/

de Waal, F. B. M., and Lanting, F. (1997). *Bonobo: The Forgotten Ape*. Berkeley, CA: University of California Press.

Diamond, J. (1986). Variation in human testis size. *Nature*, 320: 488.

Dixson, A. F. (1998). *Primate Sexuality: Comparative Studies of the Prosimians, Monkeys, Apes, and Human Beings*. New York: Oxford University Press.

Dixson, A. F. (2009). *Sexual Selection and the origins of Human Mating Systems*. Oxford: Oxford University Press.

Dixson, A. F., and Mundy, N. I. (1994). Sexual Behavior, Sexual Swelling, and Penile Evolution in Chimpanzee (*Pan troglodytes*). *Archives of Sexual Behavior*, 23(3): 267-280.

Doidge, N. (2008). *The Brain That Changes Itself: Stories of Personal Triumph from the Frontiers of Brain Science*. London: Penguin.

Donaldson, Z. R., *et al.* (2008). Evolution of a behavior-linked microsatellite-containing element in the 5' flanking region of the primate *AVPR1A* gene. *BMC Evolutionary Biology*, 8: 180.

Doran-Sheehy, D. M., Fernandez, D., and Borries, C. (2009). The strategic use of sex in wild female western gorillas. *American Journal of Primatology*, 71(12): 1011–120.

Dorus, S., Evans, P. D., Wyckoff, G. J., Choi, S. S., and Lahn, B. T. (2004). Rate of molecular evolution of the seminal protein gene SEMG2 correlates with levels of female promiscuity. *Nature Genetics*, 36(12): 1326-1329.

Dunbar, R. I. M. (1992). Neocortex size as a constraint on group size in primates. *Journal of Human Evolution*, 22: 469-493.

Dunbar, R. I. M. (1993). Coevolution of neocortical size, group size and language in humans. *Behavioral and Brain Sciences*, 16(4): 681-735.

Dunbar, R. I. M. (2010). Deacon's Dilemma: The Problem of Pair-bonding in Human Evolution. In R. Dunbar, C. Gamble, and J Gowlett (Eds.), *Social Brain, Distributed Minds*. Oxford: Oxford University Press.

Eberhard, W. G. (1996). *Female Control: Sexual Selection by Cryptic Female Choice*. Princeton NJ: Princeton University Press.

Edgerton, R. B. (1992). *Sick Societies: Challenging the Myth of Primitive Harmony*. New York: The Free Press.

Edvardsson, M. (2007). Female *Callosobruchus maculatus* mate when they are thirsty: resource-rich ejaculates as mating effort in a beetle. *Animal Behaviour*, 74(2): 183-188.

Edwards, T. C., and Collopy, M. W. (1983). Obligate and facultative brood reduction in eagles: an examination of factors that influence fratricide. *Auk*, 100: 630-635.

Elgar, M. A. (2005). Polyandry, Sperm Competition, and Sexual Conflict. In J. J. Bolhuis and L.-A. Giraldeau (Eds.) *The Behavior of Animals: Mechanisms, function, and evolution*. Oxford: Blackwell.

Emlen, S. T., and Wrege P. H. (1986). Forced copulations and intra-specific parasitism:Two costs of social living in the white-fronted bee-eater. *Ethology*, 71(1): 2-29.

Erikson, P. (2002). Several Fathers in One's Cap: Polyandrous Conception among the Panoan Matis (Amazonas, Brazi). In S. Beckerman and P. Valentine (Eds.), *Cultures of Multiple Fathers: The Theory and Practice of Partible Paternity in Lowland South America*. Gainesville, FL: University Press of Florida.

Fawcett, K., and Muhumuza, G. (2000). Death of a wild chimpanzee community member: possible outcome of intense sexual competition. *Amer. J. Primatol.* 51: 243-247.

Finkel, M. (2009). The Hadza. *National Geographic*. Retrieved July 31, 2001 from http://ngm.nationalgeographic.com/2009/hadza/finkel-text

Fisher, H. E. (1992). *Anatomy of Love.* New York: Fawcett Columbine.
Foerster, K., Coulson, T., Sheldon, B.C., Pemberton, J.M., Clutton-Brock, T.H., and Kruuk, L.E.B. (2007). Sexually antagonistic genetic variation for fitness in red deer. *Nature*, 447(28): 1107-1110.
Forbes, S. (2005). *A Natural History of Families.* Princeton, NJ: Princeton University Press.
Fox, C. A., Meldrum, S. J., and Watson, B. W. (1973). Continuous measurement by radio-telemetry of vaginal pH during human coitus. *J Reprod Fertil* 33: 69–75.
Francis, R. C. (2004). *Why Men Won't Ask For Directions: The Seductions Of Sociobiology.* Princeton, NJ: Princeton University Press.
Freund, M. (1963). Effect of frequency of emission on semen output and an estimate of daily sperm production in man. *Journal of Reproduction and Fertility*, 6: 269-286.
Furuichi, T. (1987). Sexual swelling, receptivity, and grouping of wild pygmy chimpanzee females at Wamba, Zaire. *Primates,* 28: 309-318.
Furuichi, T., and Hashimoto, C. (2002). Why female bonobos have a lower copulation rate during estrus than chimpanzees. In C. Boesch, G. Hohmann, and L. F. Marchant (Eds.), *Behavioural Diversity in Chimpanzees and Bonobos.* Cambridge: Cambridge University Press.
Furuichi, T., and Hashimoto, C. (2004). Sex differences in copulation attempts in wild bonobos at Wamba. *Primates,* 45(1): 59-62.
Gagneux, P., Boesch, C., and Woodruff, D. S. (1999). Female reproductive strategies, paternity and community structure in wild West African chimpanzees. *Animal behaviour,* 57(1): 19-32.
Gagneux, P., Woodruff, D. S., and Boesch, C. (1997). Furtive mating in female chimpanzees. *Nature,* 387: 358–359.
Gagneux, P., Woodruff, D. S., and Boesch, C. (2001). retraction: Furtive mating in female chimpanzees. *Nature,* 414: 508.
Gavrilets, S. (2000). Rapid evolution of reproductive barriers driven by sexual conflict. *Nature*, 403: 886-889.
Gerloff, U., Hartung, B., Fruth, B., Hohmann, G., and Tautz, D. (1999). Intracommunity relationships, dispersal pattern and paternity success in a wild living community of bonobos (*Pan paniscus*) determined from DNA analysis of faecal samples. *Proc. R. Soc. B.*, 266(1424): 1189–1195.
Gluckman, M. (1963). *Custom and Conflict in Africa.* Oxford: Basil Blackwell.

Goldberg, S. (2003). *Why Men Rule: A Theory of Male Dominance.* Chicago: Open Court.

Goodenough, U. (1998). *The Sacred Depths of Nature.* New York: Oxford University Press.

Gowaty P. A., Plissner, J. H., and Williams, T. G. (1989). Behavioural correlates of uncertain parentage: mate guarding and nest guarding by eastern bluebirds, *Sialia sialis. Animal Behaviour*, 38(2): 272-284.

Greenwood, P. J. (1980). Mating Systems, philopatry, and dispersal in birds and mammals. *Animal Behavior,* 28(4): 1140–1162.

Gregor, T. (1985). *Anxious Pleasures: The Sexual Lives of an Amazonian People.* Chicago: University of Chicago Press.

Gurven, M, and Kaplan, H. (2007). Longevity Among Hunter-Gatherers: A Cross-Cultural Examination. *Population and Development Review*, 33(2): 321-365.

Gwynne, D. T. (2003). Mating Behaviors. In V. H. Resh, and R. T.Carde (Eds.), *Encyclopedia of Insects.* San Diego, CA: Academic Press.

Hammock, E. A. D., and Young, L. J. (2005). Microsatellite Instability Generates Diversity in Brain and Sociobehavioral Traits. *Science,* 308: 1630–1634.

Hansson, B., Bensch, S., and Hasselquist, D. (1997). Infanticide in great reed warblers: secondary females destroy eggs of primary females. *Animal Behaviour,* 54(2): 297-304.

Hardling, R., Gosden, T., and Aguilee, R. (2008). Male mating constraints affect mutual mate choice: prudent male courting and sperm-limited females. *American Naturalist,* 172(2): 259-271.

Hashimoto, C., and Furuichi, T. (1996). Social Role and Development of Noncopulatory Sexual Behavior of Wild Bonobos. In Wrangham, R. W., McGrew, W. C., de Waal, F. B. M. & Heltne, P. G. (Eds.), *Chimpanzee Cultures.* Cambridge, MA: Harvard University Press.

Hashimoto, C., and Furuichi, T. (2006). Comparison of behavioural sequence of copulation between chimpanzees and bonobos. *Primates,* 47(1): 51-55.

Hausfater, G., and Hrdy, S. (Eds.). (1984). *Infanticide: Comparative and Evolutionary Perspectives.* New York: Aldine Publishing Co.

Heimen, D. (1997). Paternal uncertainty and ritual kinship among the Warao of the Orinoco Delta. *49 Congreso Internacional del Americanistas (ICA).* Retrieved August 7, 2011 from http://www.naya.org.ar/congresos/contenido/49CAI/Heimen.htm

Heinen, H. D., and Wilbert, W. (2002). Paternal Uncertainty and Ritual Kinship among the Warao. In S. Beckerman and P. Valentine (Eds.), *Cultures of Multiple Fathers: The Theory and Practice of Partible Paternity in Lowland South America.* Gainesville, FL: University Press of Florida.

Hennigh, L. (1970). Functions and Limitations of Alaskan Eskimo Wife Trading. *Arctic,* 23(1).

Hewlett, B. L., and Hewlett, B. S. (2008). A Biosocial Approach to Sex, Love, and Intimacy in Central African Foragers and Farmers. In W. R. Jankowiak *Intimacies: Love and Sex Across Cultures.* New York, NY: Columbia University Press.

Hewlett, B. S. (1991). *Intimate Fathers: The Nature and Context of Aka Pygmy Paternal Infant Care.* Ann Arbor, MI: University of Michigan Press.

Hewlett, B. S., and Hewlett, B. L. (2010). Sex and searching for children among Aka foragers and Ngandu farmers of Central Africa. *African Study Monographs,* 31(3):107-125.

Hill, K., Boesch, C., Goodall, J., Pusey, A., Williams, J., and Wrangham, R. (2001). Mortality rates among wild chimpanzees. *Journal of Human Evolution,* 40(5): 437-450.

Hill, K., Hurtado, A. M., and Walker, R. S. (2007). High adult mortality among Hiwi hunter-gatherers: Implications for human evolution. *Journal of Human Evolution,* 52(4): 443-454.

Hill, W. C. O., and Matthews, L. H. (1949). The male external genitalia of the gorilla, with remarks on the *os penis* of other Hominoidea. *Proceedings of the Zoological Society of London*, 119: 363-378.

Hohmann, G., and Fruth, B. (2000). Use and function of genital contacts among female bonobos. *Animal Behaviour,* 60(1): 107-120.

Hohmann, G., and Fruth, B. (2002). Dynamics in social organization of bonobos (*Pan paniscus*). In C. Boesch, G. Hohmann, and L. F. Marchant (Eds.), *Behavioural Diversity in Chimpanzees and Bonobos.* Cambridge: Cambridge University Press.

Holland, B., and Rice, W. R. (1999). Experimental removal of sexual selection reverses inter sexual antagonistic coevolution and removes a reproductive load. *Proceedings National Academy of Sciences (USA), 96: 5083-5088.*

Holmberg, A., R. (1969). *Nomads of the Long Bow: The Siriono of Eastern Bolivia.* New York: The Natural History Press.

Hosken, D. J., and Stockley, P. (2004). Sexual selection and genital evolution. *Trends in Ecology and Evolution*, 19(2): 87–93.

House, C. M., and Lewis, Z. (2007). Genital Evolution: Blurring the Battle Lines between the Sexes. *Current Biology*, 17(23): R1013-1014.

Hrdy, S. B. (1986). Empathy, Polyandry, and the Myth of the Coy Female. In R. Bleier (Ed.), *Feminist Approaches to Science*. Elmsford, NY: Pergamon Press.

Hrdy, S. B. (1996). Raising Darwin's Consciousness: Female sexuality and the pre-hominid origins of patriarchy. *Human Nature*, 8(1): 1-49.

Hrdy, S. B. (1999a). *Mother Nature: Natural Selection and the Female of the Species*. London: Chatto & Windus.

Hrdy, S. B. (1999b). *The Woman that Never Evolved*. Cambridge, MA: Harvard University Press.

Hrdy, S. B. (2000). The Optimal Number of Fathers: Evolution, Demography, and History in the Shaping of Female Mate Preferences. *Annals of the New York Academy of Sciences*, 907: 75-96.

Hrdy, S. B. (2009). *Mothers and Others: The Evolutionary Origins of Mutual Understanding*. Cambridge, MA: Harvard University Press.

Hua, C. (2001). *A Society Without Fathers or Husbands: The Na of China*. New York: Zone Books.

Hughes, J. F., Skaletsky, H., Pyntikova, T., *et al.* (2005). Conservation of Y-linked genes during human evolution revealed by comparative sequencing in chimpanzee. *Nature*, 437: 100-103.

Idani, G. (1991). Social relationships between immigrant and resident bonobo (Pan paniscus) females at Wamba. *Folia Primatologica*, 57(2): 83-95.

Innocenti, P., and Morrow, E. H. (2010).The sexually antagonistic genes of *Drosophila melanogaster*. *PLoS Biol*, 8(3): e1000335.

Jaing, X., Wang, Y., and Wang, Q. (1999). Coexistence of monogamy and polygyny in black-created gibbon. *Primates*, 40(4): 607-611.

Jensen-Seaman, M. I., and Li, W. H. (2003). Evolution of the Hominoid Semenogelin Genes, the Major Proteins of Ejaculated Semen. *Journal of Molecular Evolution*, 57: 261-270.

Jessop, T. S., Chan, R., and Stuart-Fox, D. (2009). Sex steroid correlates of female-specific colouration, behaviour and reproductive state in Lake Eyre dragon lizards, *Ctenophorus maculosus*. *J Comp Physiol A Sens Neural Behav Physiol.*, 195(7): 619-30.

Johnson, A. W., and Earle, T. (2000). *The Evolution of Human Societies: From Foraging Group to Agrarian State.* Stanford, CA: Stanford University Press.

Jolly, A. (1999). *Lucy's Legacy: sex and intelligence in human evolution.* Cambridge, MA: Harvard University Press.

Jones, A. G., Walker, D., and Avise, J. C. (2001). Genetic Evidence for Extreme Polyandry and Extraordinary Sex-Role Reversal in a Pipefish. *Proceedings: Biological Sciences,* 268(1485): 2531-2535.

Jones, S. (2002). *Y: The Descent of Men.* London: Little, Brown.

Judson, O. (2002). *Dr Tatiana's Sex Advice to All Creation: The Definitive Guide To The Evolutionary Biology Of Sex.* London: Chatto & Windus.

Kano, T. (1989). The Sexual Behavior of Pygmy Chimpanzees. In P. G. Heltne and L. A. Marquardt (eds), *Understanding Chimpanzees.* Cambridge, MA: Harvard University Press.

Kano, T. (1992). *The Last Ape: Pygmy Chimpanzee Behavior and Ecology.* Palo, Alto, CA: Stanford University Press.

Kano, T. (1996). Male rank order and copulation rate in a unit-group of bonobos at Wamba, Zaire. In W. C. McGrew, L. F. Marchant, and T. Nishida (Eds.), *Great Ape Societies.* Cambridge: Cambridge University Press.

Kelly, R. L. (2007). *The Foraging Spectrum: Diversity in Hunter-Gatherer Lifestyles.* Clinton Corners, NY: Percheron Press.

Kendrick, K. M., Hinton, M. R., Atkins, K., Haupt, M. A., and Skinner, J. D. (1998). Mothers determine sexual preferences. *Nature,* 395: 229-230.

Kesinger, K. M. (2002). The Dilemmas of Co-Paternity in Cashinahua Society. In S. Beckerman and P. Valentine (Eds.), *Cultures of Multiple Fathers: The Theory and Practice of Partible Paternity in Lowland South America.* Gainesville, FL: University Press of Florida.

King, B. J. (2004). *The Dynamic Dance : Nonvocal Communication in African Great Apes.* Cambridge, MA: Harvard University Press.

Kingan, S. B., Tatar, M., and Rand, D. M. (2003). Reduced polymorphism in the chimpanzee semen coagulating protein, Semenogein I. *Journal of Molecular Evolution,* 57: 159-169.

Knauft, B. B. (1991). Violence and Sociality in Human Evolution. *Current Anthropology,* 32: 391-428.

Knott, C. D. (2009). Orangutans: Sexual Coercion without Sexual Violence. In M. N. Muller and R, Wrangham (Eds.), *Sexual Coercion in Primates and Humans: An evolutionary perspective on male aggression against females.* Cambridge, MA: Harvard University Press.

Krokene, C., Rigstad, K., Dale, M., and Lifjeld, J. T. (1998). The function of extrapair paternity in blue tits and great tits: good genes or fertility insurance? *Behavioral Ecology,* 9(6): 649-656.

Lalueza-Fox, C., Rosas, A., et al. (2011). Genetic evidence for patrilocal mating behaviour among Neandertal groups. *PNAS,* 108(1): 250-253.

Larrick, J. W., Yost, J. A., Kapan, J., King, G., and Mayhall, J. (1979). Patterns of health and disease among the Waorani Indians of eastern Ecuador. *Medical Anthropology,* 3(2): 147-189.

Lea, V. (2002). Multiple Paternity among the Mebengokre (Kayapo, Je) of Central Brazil. In S. Beckerman and P. Valentine (Eds.), *Cultures of Multiple Fathers: The Theory and Practice of Partible Paternity in Lowland South America.* Gainesville, FL: University Press of Florida.

Lee, A., K., and Cockburn, A. (1985). *Evolutionary ecology of marsupials.* Cambridge: Cambridge University Press.

Leonard, J. L. (2005). Bateman's Principle and Simultaneous Hermaphrodites: A Paradox. *Integr. Comp. Biol.* 45: 856–873.

Leonard, J. L., and Lukowiak K. (1991). Sex and the simultaneous hermaphrodite: testing models of male-female conflict in a sea slug, *Navanax intermis* (Opisthobranchia). *Animal Behaviour*, 41(2): 255-266.

Lewis, Z., Price, T. A. R., and Weddell, N. (2008). Sperm competition, immunity, selfish genes and cancer. *Cellular and Molecular Life Sciences,* 65(20): 3241-54.

Lodge, D. (1988). *Nice Work.* London: Secker and Warburg.

Lombardi, J. (1998). *Comparative Vertebrate Reproduction.* Norwell, MA: Kluwer Academic Publishers.

Low, B. S. (2001). *Why Sex Matters. A Darwinian Look at Human Behavior.* Princeton, NJ: Princeton University Press.

Maestripieri, D. (1993). Infant kidnapping among group-living rhesus macaques: Why don't mothers rescue their infants? *Primates,* 34(2): 211-216.

Maestripieri, D., and Roney, J. R. (2005). Primate copulation calls and postcopulatory female choice. *Behavioral Ecology,* 16 (1): 106-113.

Maines, R. P. (1999). *The Technology of Orgasm: "Hysteria," the Vibrator and Women's Sexual Satisfaction.* Baltimore: John Hopkins University Press.

Majerus, M. E. N. (2003). *Sex Wars: Genes, Bacteria, and Biased Sex Ratios*. Princeton, NJ: Princeton University Press.

Marlowe, F. W. (1999). Male care and mating effort among Hadza foragers. *Behav. Ecol. Sociobiol.*, 46: 57-64.

Marlowe, F. W. (2003). A critical period for provisioning by Hadza men: Implications for pair bonding. *Evolution and Human Behavior*, 24(3): 217-229.

Marlowe, F. W. (2004a). Marital Residence among Foragers. *Current Anthropology*, 45(2): 277-284.

Marlowe, F. W. (2004b). Mate preferences among Hadza hunter-gatherers. *Human Nature, 15(4)*: 365-376.

Marshall, D. S. (1971). Sexual behavior on Mangaia. In D. S. Marshall and R. C. Suggs (Eds.), *Human Sexual Behavior*. New York & London: Basic Books.

Marson, J., Gervais, D., Meuris, S., Cooper, R. W., and Jouannet, P. (1989). Influence of ejaculatory frequency on semen characteristics in chimpanzees. *Journal of Reproduction and Fertility*, 85: 43-50.

Martin, R. D., Willner, L. A., and Dettling, A. (1994). The evolution of sexual size dimorphism in primates. In R. V. Short and E. Balaban (Eds.), *The Differences Between the Sexes*. Cambridge: Cambridge University Press.

Matsumoto-Oda, A. (1999). Female choice in the opportunistic mating of wild chimpanzees (Pan troglodytes schweinfurthii) at Mahale. *Behav. Ecol. Sociobiol.*, 46: 258-266.

Maynard-Smith, J. (1978). *The Evolution of Sex*. Cambridge: Cambridge University Press.

Maynard-Smith, J., and Szathmary, E. (1999). *The Origins of Life: From the Birth of Life to the Origins of Language*. Oxford: Oxford University Press.

Mazur, A., Booth, A., and Dabbs,, J. M. (1992). Testosterone and Chess Competition. *Social Psychology Quarterly*, Vol. 55(1): 70-77.

Mesnick, S. L. (1997). Sexual Alliances: Evidence and Evoutionary Implications. In P. A. Gowaty (Ed.), *Feminism and Evolutionary Biology: Boundaries, Intersections, and Frontiers*. New York: Chapman and Hall.

Mesoudi, A., and Laland, K. N. (2007). Culturally transmitted paternity beliefs and the evolution of human mating behavior. *Proc. R. Soc. Biology*, 274(1615): 1273-1278.

Mock, D. W., Drummond, H., and Stinson C. H. (1990). Avian Siblicide. *American Scientist*, 78(5): 438-449.

Moelike, C. W. (2001). The first case of homosexual necrophilia in the mallard *Anas platyrhynchos* (Aves: Anatidae). *DEINSEA,* 8: 243-247.

Morris, D. (1981). *The Soccer Tribe.* London: Jonathan Cape.

Muller, M. N., Kahlenberg, S. M., and Wrangham, R. W. (2009). Male Aggression against Females and Sexual Coercion in Chimpanzees. In M. N. Muller and R. W. Wrangham (Eds.), *Sexual Coercion in Primates and Humans: An Evolutionary perspective on Male Aggression Against Females.* Cambridge, MA: Harvard University Press.

Namu, Y. E., and Mathieu, C. (2004). *Leaving Mother Lake: A Girlhood at the Edge of the World.* New York: Back Bay Books.

Nielsen, R., Bustamante, C., Clark, A. G., Glanowski, S., Sackton, T. B., et al. (2005). A Scan for Positively Selected Genes in the Genomes of Humans and Chimpanzees. *PLoS Biol.,* 3(6): e170.

Nunn, C. L., Gittleman, J. L., and Antonovics, J. (2000). Promiscuity and the Primate Immune System. *Science,* 290 (5494): 1168-1170.

Olsen, D. A. (1996). *Music of the Warao of Venezuela: song people of the rainforest.* Gainesville, FL: University Press of Florida.

Olsson, M. (1995). Forced Copulation and Costly Female Resistance Behavior in the Lake Eyre Dragon, *Ctenophorus maculosus. Herpetologica,* 51(1): 19-24.

Parker, I. (2007). Swingers: bonobos are celebrated as peace-loving, matriarchal, and sexually liberated. Are they? *The New Yorker,* Jul 30: 48-61.

Parker, G.A., Baker, R.R., and Smith, V.G.F. (1972). The origin and evolution of gamete dimorphism and the male-female phenomenon. *Journal of Theoretical Biology,* 36(3): 529-553.

Paul, A., Preuschoft, S., and van Schaik, C. P. (2000). The other side of the coin: infanticide and the evolution of affiliative male-infant interactions in Old World primates. In C. P. Van Schaik, and C. H. Janson (Eds.), *Infanticide by Males and its implications.* Cambridge: Cambridge University Press.

Pawlowski, B., and Grabarczyk, M. (2003). Center of body mass and the evolution of female bodyshape. *Am. J. Hum. Biol.,* 15: 144–50.

Peluso, D. M., and Boster, J. S. (2002). Partible Parentage and Social Networks among the Ese Eja. In S. Beckerman and P. Valentine (Eds.), *Cultures of Multiple Fathers: The Theory and Practice of Partible Paternity in Lowland South America.* Gainesville, FL: University Press of Florida.

Pennington, R. (2001). Hunter-gatherer demography. In C. Panter-Brick, R. H. Layton, and P. Rowley-Conwy (Eds.), *Hunter-Gatherers: An Interdisciplinary Perspective.* Cambridge: Cambridge University Press.

Perry, G. H., Tito, R. Y., and Verrelli, B. C. (2007). The evolutionary history of human and chimpanzee Y-chromosome gene loss. *Molecular Biology and Evolution,* 24: 853-859.

Pitnick, S, Miller, G. T., Reagan, J., and Holland, B. (2001). Males' evolutionary responses to experimental removal of sexual selection. *Proc. R. Soc. B.,* 268(1471): 1071-1080.

Pitnick, S., Jones, K. E., and Wilkinson, G. S. (2006). Mating System and Brain Size in Bats. *Proc. R. Soc. B.,* 273(1587): 719-724.

Pochron, S., and Wright, P. (2002). Dynamics of testis size compensates for variation in male body size. *Evolutionary Ecology Research,* 4: 577-585.

Pollock, D. (2002). Partible Paternity and Multiple Paternity among the Kulina. In S. Beckerman and P. Valentine (Eds.), *Cultures of Multiple Fathers: The Theory and Practice of Partible Paternity in Lowland South America.* Gainesville, FL: University Press of Florida.

Prager, E. (2011). *Sex, Drugs, and Sea Slime: The Oceans' Oddest Creatures and Why They Matter.* Chicago: University of Chicago Press.

Prasad, N. G., Bedhomme, S., Day, T., and Chippindale, A. K. (2007). An evolutionary cost of separate genders revealed by male-limited expression. *Am. Nat.* 169(1): 29-37.

Pruetz, J. D., and Bertolani, P. (2007). Savanna Chimpanzees, *Pan troglodytes verus,* Hunt with Tools. *Current Biology,* 17(5): 412-417.

Pusey, A. E. (2001). Of Genes and Apes. In F.M. de Waal (Ed.), *Tree of Origin: What Primate Behavior Can Tell Us about Human Social Evolution.* Cambridge, MA: Harvard University Press.

Puts, D. A. (2010). Beauty and the beast: Mechanisms of sexual selection in humans. *Evolution and Human Behavior,* 31: 157-175.

Raffaele, P. (2010). *Among the Great Apes: Adventures on the Trail of Our Closest Relatives.* New York, NY: Smithsonian Books.

Ramm, S. A., *et al.* (2008). Sexual Selection and the Adaptive Evolution of Mammalian Ejaculate Proteins. *Molecular Biology and Evolution,* 25(1): 207-219.

Randerson, J. P., and Hurst, L. D. (2001). A comparative test of a theory for the evolution of anisogamy. *Proc. R. Soc. B*, 268: 879-884.

Reynolds, J. D., Colwell, M. A., and Cooke, F. (1986). Sexual selection and spring arrival times of red-necked and Wilson's phalaropes. *Behavioral Ecology and Sociobiology*, 18(4): 303-310.

Řezáč, M. (2009). The spider *Harpactea sadistica*: co-evolution of traumatic insemination and complex female genital morphology in spiders. *Proc. R. Soc. B*, 276: 2697–2701.

Rice, W. R. (1996). Sexually antagonistic male adaptation triggered by experimental arrest of female evolution. *Nature*, 381: 232-234.

Rice, W. R., and Chippindale, A. K. (2001), Intersexual ontogenetic conflict. *Journal of Evolutionary Biology*, 14: 685–693.

Ridley, Matt (1994). *The Red Queen: Sex and the Evolution of Human Nature*. London: Penguin.

Ridley, Mark (2000). *Mendel's Demon: Gene justice and the complexity of life.* London: Weidenfield and Nicolson.

Rodriguez, A. M., and Monterrey, N. S. (2002). A Comparative Analysis of Paternity among the Piaroa and the Ye'kwana of the Guayana Region of Venezuela. In S. Beckerman and P. Valentine (Eds.), *Cultures of Multiple Fathers: The Theory and Practice of Partible Paternity in Lowland South America.* Gainesville, FL: University Press of Florida.

Rogers, A. R., Iltis, D., and Wooding, S. (2004). Genetic variation at the *MC1R* locus and the time since loss of human body hair. *Current Anthropology*, 45:105-108.

Rogers, D. W., and Chase, R. (2001). Dart receipt promotes sperm storage in the garden snail Helix aspersa. *Behavioral Ecology and Sociobiology*, 50: 122-127.

Ross, C. N., French, J. A., and Orti, G. (2007). Germ-line chimerism and paternal care in marmosets (*Callithrix kuhlii*). *PNAS*, 104(15): 6278-6282.

Rosso, L., *et al.* (2008). Mating system and *avpr1a* promoter variation in primates. *Biol. Lett.*, 4: 375-378.

Rouse, G. W., Goffredi, S. K., and Vrijenhoek, R. C. (2004). *Osedax*: Bone-Eating Marine Worms with Dwarf Males. *Science*, 305(5684): 668-671.

Rousseau, J-J. Transl. Foxley, B. (1993). *Emile*. London: Everyman.

Ryan, C., and Jethá, C. (2010). *Sex at Dawn: The Prehistoric Origins of Modern Sexuality.* New York: HarperCollins.

Ryan, C., and Jethá, C. (2011). *Sex at Dawn: How We Mate, Why We Stray, and What It Means for Modern Relationships.* New York: HarperCollins.

Ryne, C. (2009). Homosexual interactions in bed bugs: alarm pheromones as male recognition signals. *Animal Behaviour,* 78(6): 1471-1475.

Sakaluk, S. K., Campbell, M. T. H., Clark, A. P., Johnson, J. C., and Keorpes, P. A. (2004). Hemolymph loss during nuptial feeding constrains male mating success in sagebrush crickets. *Behavioral Ecology,* 15(5): 845-849.

Sandell, M. I. (1998). Female Aggression and the Maintenance of Monogamy: Female Behaviour Predicts Male Mating Status in European Starlings. *Proc. R. Soc. B.,* 265(1403): 1307-1311

Sandell, M. I., and Smith, H. G. (1996). Already Mated Females Constrain Male Mating Success in the European Starling. *Proc. R. Soc. B.,* 263(1371): 743-747.

Schaller, F., *et al.* (2010).Y Chromosomal Variation Tracks the Evolution of Mating Systems in Chimpanzee and Bonobo. *PLoS ONE,* 5(9): e12482.

Schlaberg, R., Choe, D. J., Brown, K. R, Thaker, H. M., and Singh, I. R. (2009). XMRV is present in malignant prostatic epithelium and is associated with prostate cancer, especially high-grade tumors. *Proc Natl Acad Sci U S A.,* 106(38): 16351–16356.

Schlegel, A. (1995). The cultural management of adolescent sexuality. In P. R. Abramson and S. D. Pinkerton (Eds.), *Sexual Nature/Sexual Culture.* Chicago: University of Chicago Press.

Shih, C-K. (2010). *Quest for Harmony: The Moso Traditions of Sexual Union and Family Life.* Stanford, CA: Stanford University Press.

Shih, C-K, and Jenike, M. R. (2002). A Cultural-Historical Perspective on the Depressed Fertility among the Matrilineal Moso in Southwest China. *Human Ecology,* 30(1): 21-47.

Shine, R., Langkilde, T., and Mason, R. T. (2003). Cryptic Forcible Insemination: Male Snakes Exploit Female Physiology, Anatomy, and Behavior to Obtain Coercive Matings. *The American Naturalist,* 162(5): 653-667.

Shostak, M. (1990). *Nisa: The Life and Words of a !Kung Woman.* London: Earthscan.

Shuster, S. M., and Wade, M. J. (2003). *Mating Systems and Strategies.* Princeton: Princeton University Press.

Silk, J. B. (1980). Kidnapping and female competition among captive bonnet macaques. *Primates,* 21(1): 100-110.

Skau, P. A., and Folstad, I. (2004). Does immunity regulate ejaculate quality and fertility in humans? *Behavioral Ecology*, 16(2):410-416.

Small, M. F. (1993). *Female Choices: Sexual Behavior of Female Primates.* Ithaca, NY: Cornell University Press.

Smuts, B. B., and Gubernick, D. J. (1992). Male-Infant Relationships in Nonhuman Primates: Paternal Investment or mating Effort? In B. S. Hewlett (Ed.), *Father-Child Relations: Cultural and Biosocial Contexts.* New York: Aldine De Gruyter.

Soronen, P., Laiti, M., Torn, S., Harkonen, P., Patrikainen, L., Li, Y., et al. (2004). Sex steroid hormone metabolism and prostate cancer. *Journal of Steroid Biochemistry and Molecular Biology*, 92(4): 281-286.

Sparks, J. (1999). *Battle of the Sexes: The Natural History of Sex.* London: BBC Books.

Stanford, C. (1998). The Social Behavior of Chimpanzees and Bonobos: Empirical Evidence and Shifting Assumptions. *Current Anthropology*, 39(4): 399-420.

Stanford, C. (2000). The Brutal Ape vs. the Sexy Ape? *American Scientist*, 88(2): 110-112.

Stanford, C. (2001). *Significant Others: The Ape - Human Continuum and the Quest for Human Nature.* New York: Basic Books.

Stumpf, R. M., and Boesch, C. (2005). Does promiscuous mating preclude female choice? Female sexual strategies in chimpanzees (Pan troglodytes verus) of the Taï National Park, Côte d'Ivoire. *Behav. Ecol. Sociobiol.*, 57: 511-524.

Sturma, M. (2002). *South Sea Maidens: Western Fantasy and Sexual Politics in the South Pacific.* New York: Praeger.

Suarez, S. S., and Pacey, A. A. (2006). Sperm transport in the female reproductive tract. *Hum. Reprod. Update*, 12(1): 23-37.

Summers, K., and Crespi, B. (2007). The androgen receptor and prostate cancer: A role for sexual selection and sexual conflict? *Medical Hypotheses*, 70(2): 435-443.

Surbeck, M., Deschner, T., Schubert, G., Weltring, A., and Hohmann, G. (2012). Mate competition, testosterone and intersexual relationships in bonobos, *Pan paniscus. Animal Behaviour*, 83(3): 659-669.

Surbeck, M., and Hohmann, G. (2008). Primate hunting by bonobos at LuiKotale, Salonga National Park. *Current Biology*, 18(19): R906-7.

Surbeck, M., Mundry, R., and Hohmann, G. (2010). Mothers matter! Maternal support, dominance status and mating success in male bonobos (*Pan paniscus*). *Proc. R. Soc. B., 278: 590–598.*
Takahata, Y., Ihobe, H., and Idani G. (1996). Comparing copulations of chimpanzees and bonobos: do females exhibit proceptivity or receptivity? In W. C. McGrew, L. F. Marchant, and T. Nishida (Eds.), *Great Ape Societies.* Cambridge: Cambridge University Press.
Tang-Martinez, Z., and Brandt Ryder, T. (2005). The Problem with Paradigms: Bateman's Worldview as a Case Study. *Integrative and Comparative Biology,* 45(5): 821-830.
Townsend, S. W., Deschner, T., and Zuberbühler, K. (2008) Female Chimpanzees Use Copulation Calls Flexibly to Prevent Social Competition. *PLoS ONE* 3(6): e2431.
Trivers, R., L. (1972). Parental investment and sexual selection. In B. Campbell (Ed.), *Sexual Selection and the Descent of Man.* Chicago: Aldine.
Valentine, P. (2002). Fathers that Never Exist: Exclusion of the Role of Shared Father among the Curripaco of the Northwest Amazon. In S. Beckerman and P. Valentine (Eds.), *Cultures of Multiple Fathers: The Theory and Practice of Partible Paternity in Lowland South America.* Gainesville, FL: University Press of Florida.
Valera, F. Hoi, H., and Kristin, A. (2003). Male shrikes punish unfaithful females. *Behavioral Ecology,* 14(3): 403-408.
van der Meij, L., Buunk, A. P., van de Sande, J. P., and Salvador, A. (2008). The presence of a woman increases testosterone in aggressive dominant men. *Hormones and Behavior,* 54: 640–644.
van Schaik, C. P., and Janson, C. H. (Eds.). *Infanticide by Males and its implications.* Cambridge: Cambridge University Press.
Veiga, J. P. (1990). Sexual Conflict in the House Sparrow: Interference between Polygynously Mated Females versus Asymmetric Male Investment. *Behavioral Ecology and Sociobiology,* 27(5): 345-350.
Veiga, J. P. (1993). Prospective infanticide and ovulation retardation in free-living house sparrows. *Animal Behaviour,* 45(1): 43-46.
Vickers, W. T. (2002). Sexual Theory, Behavior, and Paternity among the Siona and Secoya Indians of Eastern Ecuador. In S. Beckerman and P. Valentine (Eds.), *Cultures of Multiple Fathers: The Theory and Practice of Partible Paternity in Lowland South America.* Gainesville, FL: University Press of Florida.
Vigilant, L. *et al.* (2001). Paternity and relatedness in wild chimpanzee communities. *PNAS,* 98(23): 12890-12895.

Wade, M. J., and Shuster, S. M. (2005). Don't Throw Bateman Out with the Bathwater! *Integrative and Comparative Biology,* 45(5): 945-51.

Walker, M. (2009). Bees fight to death over females. Retrieved July 7, 2011 from *BBC Earth News,* http://news.bbc.co.uk/earth/hi/earth_news/newsid_8354000/8354788.stm

Walker, R. S, Flinn, M. V., and Hill, K.R. (2010). Evolutionary history of partible paternity in lowland South America. *PNAS* 107(45): 19195-19200.

Walker, R. S, Hill, K.R, Flinn, M. V, and Ellsworth R. M. (2011). Evolutionary History of Hunter-Gatherer Marriage Practices. *PLoS ONE,* 6(4): e19066.

Wallen, K. (2001). Sex and Context: Hormones and Primate Sexual Motivation. *Hormones and Behavior,* 40: 339-357.

Watts, D. P. (2004). Intracommunity Coalitionary Killing of an Adult Male Chimpanzee at Ngogo, Kibale National Park, Uganda. *International Journal of Primatology,* 25(3): 507-521.

White, F. J. (1992). Eros of the Apes. *BBC Wildlife Magazine,* 10(8): 38-47.

White, F. J. (1996a). Comparative socio-ecology of *Pan paniscus.* In W. C. McGrew, L. F. Marchant, and T. Nishida (Eds.), *Great Ape Societies.* Cambridge: Cambridge University Press.

White, F. J. (1996b). *Pan paniscus* 1973 to 1996: Twenty-Three Years of Field Research. *Evolutionary Anthropology,* 5(1): 11-17.

White, F. J., Wood, K. D., and Merrill M. Y. (1998). Comments. In The Social Behavior of Chimpanzees and Bonobos: Empirical Evidence and Shifting Assumptions. *Current Anthropology,* 39(4): 399-420.

White, T. D., *et al.* (2009). *Ardipithecus ramidus* and the Paleobiology of Early Hominids. *Science,* 326(5949): 75-86.

Whitten, P. L. (1987). Infants and Adult Males. In B. B. Smuts, D. L. Cheney, R. M. Seyfarth, R. W. Wrangham, and T. T. Struhsaker (Eds.), *Primate Societies.* Chicago: University of Chicago Press.

Williams, D. M. (1983). Mate choice in the Mallard. In P. Bateson (Ed.), *Mate Choice.* Cambridge: Cambridge University Press.

Williams, G. C. (1975). *Sex and Evolution.* Princeton: Princeton University Press.

Wilkinson, G. S. (1985). The social organization of the common vampire bat: II. Mating System, Genetic Structure, and Relatedness. *Behavioral Ecology and Sociobiology,* 17(2): 123-134.

Wilson, E. O. (1975). *Sociobiology: The New Synthesis.* Cambridge, MA: Harvard University Press.
Wilson, E. O. (1978). *On Human Nature.* Cambridge, MA: Harvard University Press.
Wingfield, J. C., Hegner, R. E., Dufty, A. M., and Ball, G. F. (1990). "The "Challenge Hypothesis": Theoretical Implications for Patterns of Testosterone Secretion, Mating Systems, and Breeding Strategies". *The American Naturalist,* 136(6): 829–846.
Wrangham, R. (2002). The cost of sexual attraction: is there a trade-off in female *Pan* between sex appeal and received coercion? In C. Boesch, G. Hohmann, and L. F. Marchant (Eds.), *Behavioural Diversity in Chimpanzees and Bonobos.* Cambridge: Cambridge University Press.
Wroblewski, E. E., Murray, C. M., Keele, B. F., Schumacher-Stankey, J. C., Hahn, B. H., and Pusey, A. E. (2009). Male dominance rank and reproductive success in chimpanzees, *Pan troglodytes schweinfurthii. Animal Behavior,* 77(4): 873-885.
Wyckoff, G. J., Wang, W., and Wu, C. (2000). Rapid evolution of male reproductive genes in the descent of man. *Nature,* 403: 304-308.
Zuk, M. (2007). *Riddled with Life: Friendly Worms, Ladybug Sex, and the Parasites That Make Us Who We Are.* Orlando: Harcourt.

INDEX

Aché, 120, 128–29, 161, 179, 187, 216, 218, 330
açia relationships, 169
adolescents, 300–302
  Canela sex restrictions, 166
adultery, 136, 137, 200, 321
aggression, 203, 214, 294
  bonobo, 87, 90, 94, 95–97, 100, 103, 105, 228, 281, 310
  chimpanzee, 88, 89, 102, 226, 235
  chimpanzee, lethal, 228
  chimpanzee, sexual, 279
Aka, 175, 194–96, 277
allomothers. *See* shared parenting
anal sex, 99, 194
anisogamy, 49, 52
antechinus, 23–24, 220, 268, 272, 304
Anthophora plumipes (bee), 45
anti-aphrodisiac, fruit fly, 19
apoptosis, 257
Argentine duck, 46
arranged marriage, 136, 144, 214, 215
assortative mating, 68
attractivity, defined, 72
Australian jewel beetle, 67
Australopithecus, 131, 143, 244

baboons, 68, 78, 81, 82, 150, 233, 244
  gelada, 75, 81, 285
  Hadza 'five-baboon' rule, 331
  hamadryas, 75, 81, 233
  male protecting own young, 74
  paternal care as mating effort, 139
  polygyny in, 75, 81, 233
Bari, 120
Barnacle Bill the Sailor, drinking song, 224
barnacles, dwarf males, 13
Bateman, A. J., 18–21, 49
Baumeister, Roy, 290, 291, 293, 301
Beach, Frank, 71, 72
Beagle, HMS, 33
Beckerman, Stephen, 112, 131, 132, 219
bedbugs, 56
Bernard, Jessie, 313–16, 317, 332
Betzig, Laura, 156
bipedalism, transport of offspring and, 222
birds, 186–89
  cloacal kiss, 59
  extra-pair paternity, 186
  siblicide, 60
bluebirds, 187

blue-headed wrasse, 54
Boesch, Christophe, 83, 88, 227, 228, 249
bonobos, 84–87, 90–108, 141, 143, 226, 228, 230, 292
   copulation calls, 280–81
   distinction between sex and "sex", 98
Bounty, HMS, 33–34
breasts, 218, 223, 282–86
Brennan, Patricia, 46
buttocks, 223, 283, 286

Cai Hua, 168, 169, 173, 185
caloric restriction, 218, 282, 288, 331
Canela, 122–26, 134, 163, 204, 205, 226, 277
Carter, C. Sue, 183, 184
Cashinahua, 112
cervix, 260, 261, 288, 289
Chapais, Bernard, 144, 235, 237, 246
Chernela, Janet, 153
chimerism, marmoset, 150
chimpanzees, 77–107, 239
   copulation calls, 279, 282
   penis size, 260
Chivers, Meredith, 291, 293
circumcision, 302
clitoris, 276
clownfish, 54
competition, 26, 61, 200, 201, 220, 223, 224, 246
   female, 88, 95, 106, 237, 245, 279, 281, 327, 329
   male, 13, 24, 56, 57, 70, 81, 127, 155, 197, 198, 199, 215, 218, 220, 301, 309, 310, 311, 319, 322
Coolidge effect, 303, 305, 307
copulatory plugs, 260, 261
   chimpanzee, 256
   honeybee, 45

Crocker, William, 121, 123, 125, 204, 205
Cronin, Helena, 185
cryptic female choice, 22, 255
Curripaco, 113, 161

Darwin, 12, 13, 14, 15, 18, 31, 34, 35, 39, 48, 56, 70, 249, 289
Davenport, William, 311, 312
Dawson's bee, 46
de Beauvoir, Simone, 2, 322
de Waal, Frans, 96, 98, 99, 100, 101, 104, 107, 301
deep sea angler fish, 16
Descent of Man, The (C. Darwin), 13, 70
Diamond, Jared, 267
digger wasp, 48
diploid cells, 50, 66
divorce, 116, 129, 161, 162, 191, 195, 208, 317, 319
Dixson, Alan F., 260, 261, 263, 267, 288, 289
DNA, 91, 92, 132, 252, 254
Dogon, 196
Dunbar, Robin, 237, 238
dunnock, 187

Edgerton, Robert, 166
egalitarianism, 142, 175, 194, 213, 215, 237, 238
egg dumping, 187
egg-trading, 29
ejaculation, 253, 265, 268, 275, 322, 328
   bonobo, 99, 101
   sperm depletion, 264
erection, 58
Erikson, Philippe, 121, 126
Ese Eja, 114
euthanasia, 216
Evolution of Human Societies, The (Johnson and Earle), 214

evolution, definition of, 14
evolutionary biology, 11, 14, 17, 18, 20, 32, 33, 36, 44, 160, 181, 185, 257
evolutionary psychology, 11, 12, 17, 36, 44
external fertilization, 57
extra-pair paternity in birds, 186, 187, 189
extra-pair sex, 136, 181, 183, 184, 186, 189, 238, 330

fat, female reproductive, 143, 222, 223, 246, 283, 284, 285
Female Choices (Small), 159
female sexuality, 18, 31, 32, 33, 276, 277, 290, 294, 295, 317, 323
   gullibility and, 290, 301, 332
   Rousseau and, 85
   situation-dependent, 137, 140, 294, 295
fertilization, conflict over control of, 29, 46, 57
fission-fusion, 226, 227, 229, 230, 231, 234, 235, 236
food in exchange for sex, 63, 64, 72, 94, 100, 105, 107, 192, 203, 207, 218, 294, 328
food, sperm as, in hermaphrodites, 28, 29
foragers. *See* hunter-gatherers
Foraging Spectrum, The (Kelly), 213
foundling hospitals, 155, 156
fruit fly, 19, 20, 69
   toxic semen, 26, 152

gangbang, 259
garden snails, 28
garter snakes, 58
gene flow, 8, 41, 42, 232, 308
gene's eye view, 7
genetic research, 91, 253, 254, 255, 256, 257, 258, 259, 269

genital echo theory, 283
genital-genital (g-g) rubbing, 93, 94, 95, 96, 99, 100, 101, 106
gibbons, 62, 73, 150, 174, 243, 247, 257
   testes, 251
Goldberg, Steven, 176, 177
Gombe, 82, 84, 88, 102
   hunting, 84
gonads, 50
good genes, 64, 69
Goodall, Jane, 82, 87
gorillas, 71, 79, 95, 246, 251, 329
   female competition, 95, 329
   infanticide by males in, 61, 74
   penis, 260, 263
   testes, 251
great reed warbler, 187
green spoon worm, *Bonellia viridis*, 55
Gregor, Thomas, 126, 127
grey langur, 73
group cohesion, 145, 232, 239
group marriage, 36, 40, 44, 156, 202, 315
group selection, 7, 9, 43

Hadza, 139, 190, 191, 216, 218, 330, 331
hamlet fish, 29
hanuman langurs, 74
haploid cells, 50, 66
harlequin bass, 29
Harpactea sadistica spider, 56
hermaphrodites, 27–30, 53, 54, 55
hierarchy
   bonobo and mating success, 95, 106
   bonobo female, 281, 308
   of attractiveness, 200
Hierarchy in the Forest, (Boehm), 215
Hill, Kim, 128, 145

Hobbes, Thomas, 84, 212
Hohmann, Gottfried, 94, 105
Holmberg, Allan R., 202, 217
Homo erectus, 78, 143, 222, 244
Homo sapiens, 1, 2, 5, 6, 222
homosexuality, 48, 146, 166, 194, 320
honeybee, 45, 46
house sparrow, 188
Hrdy, Sarah Blaffer, 73, 128, 149, 150, 151, 152, 160, 204, 205, 216, 224, 249, 250, 255, 284, 293, 294
hunter-gatherers, 140, 142, 144, 145, 158, 160, 172, 190, 193, 195, 213, 215, 216, 225, 234, 277, 278, 282, 295, 317, 333
    fission-fusion, 230
    mortality rates, 217, 218
    polygyny in, 247
hypodermic insemination. *See* traumatic insemination
hysteria, 276, 277, 296

immunological compatibility, 286
inbreeding, 212, 232, 233, 321
    avoidance, 41, 80, 89, 230
incest, 40, 41, 114, 120, 157, 212, 307, 308, 332
infanticide, 60, 61, 96, 118, 155, 217, 224
    by females, 61, 187
    by males, 60–62, 73, 74, 149, 189, 238, 294, 304, 328
infant-sharing, 149, 150
infertility, 166, 269, 284, 293
infidelity, 118, 154, 193, 198, 199, 237, 331
Insel, Thomas, 182, 183
interdependence of spouses, 133, 167, 193, 194, 196, 208, 214, 216, 238, 245, 317
intimacy, emotional, 253, 295, 323
Inuit, 165

jacanas, 61
jealousy, 114, 117, 119, 121, 123, 126, 128, 130, 142, 164, 165, 170, 195, 197, 198, 202, 204–6, 301, 303

Kayapo, 115
Kibale, 87, 89
King, Barbara J., 87
Kulina, 116–17, 167
!Kung, 162, 215, 216, 331
    Nisa, 191–94

Lake Eyre dragon lizard, 58
Larrick, James, 219
Leaving Mother Lake (Namu and Mathieu), 170
lemurs, 62
Linnaeus, Carl, 15
Lorenz, Konrad, 36
Low, Bobbi, 129
Lugu Lake Mosuo Cultural Development Association, 171, 178
Lusi, 132

Mae Enga, 197
major histocompatibility complex genes (MHC), 287
male bonding, 81, 96, 122, 123, 125, 126, 130, 133, 134, 226, 324
male parental care, 62, 140, 142, 151, 160, 186
    and lower testosterone, 311
    as mating effort, 139
male parental investment, 64, 137, 159, 160, 186, 188, 235
male philopatry, evolution of, 78–82
mallard duck, 15, 47
    rape, 47
Malthus, Thomas, 34–35, 220
mamuse relations, 163
Mangaia, 301–2
March of the Penguins, 182

Marind-anim, 166
marmosets, 62, 63, 64, 150
marriage, 32, 70, 131, 133, 144, 145, 152, 153, 155, 158, 159, 160–63, 164, 165, 167, 173, 174, 178, 181, 190, 194, 215, 216, 233, 234, 237, 239, 277, 281, 303, 306, 314–17, 329, 330, 331, 332
   and testosterone levels, 309
   Mosuo, 168, 170, 171
masturbation, 193, 195, 276
   Aka, no term for, 194
   bonobo, 99
   Canela, strictly forbidden in, 123, 125, 163, 277
mate choice, 56, 136, 139, 198, 200, 201
   bonobos, 91, 248
   female, 13, 17, 18, 22, 32, 33, 63, 67, 70, 126, 137, 159, 174, 188, 225, 255, 287, 293, 301, 328
   male, 31, 33, 159, 327
mate preference, 42, 157, 158, 191, 201, 287
mate-guarding, 75, 184–90
   by females, 327
   by males, 23, 62, 73, 75, 81, 235, 238
   prairie vole, 183
mating effort, 70, 139, 140, 160, 208
Matis, 120–21, 126
matriarchy, 127, 170, 175–79
matrilocal residence, 41, 81, 152, 157, 167, 173, 202
Mehinaku, 126–27, 147, *See also* phallic aggression
meiosis, 66
menstrual cycles, 71
Mesnick, Sarah, 188
Minangkabau, 175–77
mitochondria, 52, 53, 132, 267
monogamy, 27, 82, 137, 142, 146, 151, 152, 157, 174, 181–90, 198–200, 208, 213, 236, 238, 247, 256, 263, 264, 265, 269, 277, 308, 318, 330
   enforced in fruit flies, 26, 152
   serial, 171, 190, 324, 331
Morgan, Lewis Henry, 36–37, 39, 40, 42, 156, 157, 212
Morris, Desmond, 133
mortality, due to the evolution of sexual reproduction, 50
Mosuo, 167–75, 175, 176, 178, 181, 185, 206, 207, 329
Mother Nature (Hrdy), 149, 204, 205
multilocal residence, 144
Muria, 302–3

natural selection, 14, 11–15, 17, 20, 30, 31, 32–36, 71, 74, 139, 153, 158, 220–21, 269, 270
Navanax inermis, 29
Neanderthals, 77, 132, 222
New York Times, 88, 151, 293
New Yorker, 98
Nisa: The life and words of a !Kung woman (Shostak), 191
novelty, sexual, 293, 304, 307, 308, 313, 316–19
nuclear family, 62, 98, 144, 155, 156, 157, 165, 176, 183, 190, 193, 196, 202, 216, 229, 324
nuptial gift, 63

On the Origin of Species (C. Darwin), 12
operational sex ratio, 22
oral sex, 99, 193, 194
orangutans, 64, 65, 91, 93, 251, 254, 256, 257, 267
   rape, 64
orgasm, 92, 99, 276, 277, 275–78, 286, 287, 288, 290
   Canela, 123
   during sexual assault, 293

orgasm *(contd.)*
   Mehinaku, 126
   multiple, 275
Original Affluent Society, The
   (Sahlins), 212–14
ovarian cancer, 269
oviduct, 267
ovulation, 71, 98, 105, 142, 289
   concealed, 72, 73, 103, 284
   rarity of, 279, 288
oxytocin, 92, 93, 182

pair bonds, 68, 72–73, 80, 139, 142, 144, 143–46, 159, 160, 181–82, 184–85, 188, 198, 207, 208, 222, 229, 264, 282, 288, 317, 320, 330
Paranthropus, 131
paraphilias, 299
Parental Investment and Sexual Selection (Trivers), 158
parenting effort, 70, 133, 139, 140, 160, 208
Parker, Geoffrey, 22
paternity, 61, 62, 138, 139, 142, 144, 237, 249, 317
   confusion of, 73, 74, 138, 279, 294
   partible, 111–34, 138, 142, 153–55, 166, 277
patrilocal residence, 41, 131, 132, 152, 157, 167, 197, 233
pelvic inflammatory disease
   Marind-amin, 167
   Mosuo, 174
penduline tit, 59
penis, 57, 58, 260–63, 275
   as a plunger, 260
   comparison with other primates, 260–63
   fencing, *Pseudobiceros bedfordi*, 27
   super- and subincision of, 302
   waterfowl, 46
Pennington, Renee, 224

phalarope, 21, 22
phallic aggression, Mehinaku, 127
Piaroa, 118
pipefish, 21
Pirahã, 164
Pitcairn Island, 34
plasticity, erotic, 299, 301, 302, 314, 323
polyandry, 14, 18, 68, 181, 189, 196, 208
   convenience, 189
polygyny, 75, 78, 80, 81, 82, 130, 142, 143, 144, 178, 194, 208, 233, 236, 246, 247, 263, 264
polyspermy, 255, 256, 289
population, 217, 223, 233, 331
   founder, 272
   founder, Bonobo, 91, 106
   growth, 220–21, 224–25, 229, 231, 266
Power, Margaret, 226–28, 229
prairie vole, 92, 182–84
Primeval Kinship: How Pair-Bonding Gave Birth to Human Society (Chapais), 144, 235
proceptivity, defined, 72
prolactin, 57
promiscuity, 139, 182, 185, 200, 218, 224, 239
   as defined by Ryan and Jethá, 41
   primate immune system and, 174
prostate cancer, 268–71
Pusey, Anne, 88–90
Puts, David, 246

rape, 199, 253, 268, 293
   Inuit, of captured women, 165
   Mosuo, 171
   orangutan, 64
   punitive gang, 134
   punitive, Canela, 116, 124, 125
   punitive, Mehinaku, 127
   Siriono, 203

receptivity, defined, 72
red deer, 69, 304
repetitive microsatellite, 92, 182
reproductive tract, female, 287
   bedbug, 56
   duck, 46
   green spoon worm, 55
   Harpactea sadistica spider, 56
   newt, 58
   sperm competition and, 250, 255
reproductive tract, male, 270
reversed sex roles, 21
Rousseau, Jean-Jacques, 84–85
runaway coevolution, 255

Sahlins, Marshall, 212
Sanday, Peggy Reeves, 175–77
seahorses, 57
Secoya, 118
selfish genes, 8, 23–25, 32, 70, 199, 304, 306, 308, 313, 319, 322
semen, 166, 167, 193, 194, 255, 256, 268, 290, *See also* paternity, partible
senescence, 50
sex determination, 53
sexual coercion, 238
   bonobo, 105
   chimpanzee, 86, 102, 103
sexual conflict, 25, 30, 25–30, 207, 254, 255
   hermaphrodites, 27
   intralocus, 69
sexual dimorphism, 15, 55, 78, 80, 81, 243–47
sexual division of labour, 131, 133, 137, 142, 143, 167, 175, 177, 193, 194, 196, 202, 208, 214, 216, 222, 229, 236, 238, 245, 246
sexual harassment, 45, 65, 68, 188, 189, 238, 301
sexual interference, bonobos, 95, 248

sexual selection, 13, 16–17, 20, 22, 31, 33, 36, 38, 39, 49, 56, 57, 69, 159, 259, 268, 269, 321
   and penile morphology in primates, 261
sexual swellings, absence of in hominins, 284
sexual traits, distinction between primary and secondary, 56
sexually antagonistic co-evolution, 27
sexually transmitted disease, 64
   Mosuo infertility and, 173, 174
   primate immune system and, 174
shared parenting, 64, 149, 150, 152, 236
siamangs, 63
siblicide, 60
Siona, 118
Siriono, 164, 165, 202–3, 218, 247, 321
   abandonment of sick and elderly, 217
slipper 'limpet', *Crepidula fornicata,*, 54
sluts, 64, 130, 159, 328, 329
Small, Meredith, 159
social networks, 137, 144, 145, 160, 194, 215, 229, 232, 233, 241, 288
Society Without Fathers or Husbands, A (Cai Hua), 168
sociobiology, 36
Socio-Erotic Exchanges (S.E.Ex.), 134
South Sea Maidens: Western Fantasy and Sexual Politics in the South Pacific (Sturma), 134
sperm competition, 22, 42, 91, 103, 141, 200, 225, 232, 241, 248–68, 271, 272, 278, 279, 280, 281–82, 287, 288
sperm depletion, 31, 264
sperm midpiece, 267

sperm production, 31, 91, 256, 264–66, 269, 270, 274
spermatophore, 57, 59, 63
standard narrative, 4, 12, 181, 182, 184, 256
Stanford, Craig, 82, 83, 101
starlings, 187
stress, 220, 271
   antechinus, 23
   bonobos, 94, 100
   female primates, 238
   garter snake, 58
Sturma, Michael, 134, 135

Tahiti, 134–35
Taï chimpanzees, 86, 87, 88, 102, 104, 228, 249
   meat sharing, 82–84
   territorial attacks, 227
tamarins, 62, 63, 64, 150
testes, 50, 248, 251–52, 253, 254, 259, 264, 265, 267, 268, 271, 332
   bats, 141
   seasonal changes in sifaka, 251, 252
testosterone, 58, 220, 268, 271, 285, 299, 300, 308–11, 315, 318, 319
The Canela: Bonding Through Kinship, Ritual, and Sex (Crocker and Crocker), 204
Tinbergen, Niko, 36
titi monkey, 185
traumatic insemination, 56
Trivers, Robert L., 21, 158
Tukanoan, 153, 154, 197

vagina, 94, 260, 261
   chimpanzee, 260
   pH, 290
   reflexive sexual readiness, 293, 294
Valentine, Paul, 112

vampire bats, 140
vas deferens, 267
vasopressin, 92, 93
vasopressin receptor AVPR1A, 92, 182
Venus of Willendorf, 282
violence
   !Kung, 191, 193
   Aka, 195
   bonobo, 97, 105, 228
   chimpanzee, 87, 89, 228
   Mehinaku, 127
   Mosuo, 169
   Waorani, 219, 220
virginity, 162, 163
vocalizations, female copulatory, 277–82

Wallen, Kim, 310
Wanano. *See* Tukanoan
Waorani, 219–20
Warao, 119, 163, 164
warfare, 114, 122, 123, 126, 131, 132, 134, 136, 145, 153, 165, 190, 197, 204, 205, 214, 218, 219, 220, 225, 226, 324, 331
whalebone-eating worm, *Osedax*, 55
White, Frances, 95, 104, 105, 108
white-fronted bee-eater, 188
whores, 64, 130, 159, 328, 329
   bonobo females as, 108
Wilson, E. O., 35, 48, 146, 207, 320

X chromosome, 66

Y chromosome, 91, 258
Yang Erche Namu, 168
Yanomami, 119, 130
Ye'kwana, 117

zero-sum sex, 198–99
Zuk, Marlene, 271

Made in the USA
Lexington, KY
10 November 2014